AF251360

Integrated Circuits and Systems

Series Editor
Anantha Chandrakasan
Massachusetts Institute of Technology
Cambridge, Massachusetts

For further volumes, go to
http://www.springer.com/series/7236

Amith Singhee • Rob A. Rutenbar

Editors

Extreme Statistics in Nanoscale Memory Design

 Springer

Editors
Amith Singhee
IBM Thomas J. Watson Research Center
1101 Kitchawan Road, Route 134
Yorktown Heights, NY 10598, USA
asinghe@us.ibm.com

Rob A. Rutenbar
Department of Computer Science
University of Illinois at Urbana-Champaign
201 North Goodwin Avenue
Urbana, IL 61801, USA
rutenbar@illinois.edu

ISSN 1558–9412
ISBN 978-1-4419-6605-6   e-ISSN 978-1-4419-6606-3
DOI 10.1007/978-1-4419-6606-3
Springer New York Dordrecht Heidelberg London

Library of Congress Control Number: 2010934212

Printed on acid-free paper

Springer is part of Springer Science+Business Media (www.springer.com)

Preface

Knowledge exists: you only have to find it

VLSI design has come to an important inflection point with the appearance of large manufacturing variations as semiconductor technology has moved to 45 nm feature sizes and below. If we ignore the random variations in the manufacturing process, simulation-based design essentially becomes useless, since its predictions will be far from the reality of manufactured ICs. On the other hand, using design margins based on some traditional notion of worst-case scenarios can force us to sacrifice too much in terms of power consumption or manufacturing cost, to the extent of making the design goals even infeasible. We absolutely need to explicitly account for the statistics of this random variability, to have design margins that are accurate so that we can find the optimum balance between yield loss and design cost. This discontinuity in design processes has led many researchers to develop effective methods of *statistical design*, where the designer can simulate not just the behavior of the nominal design, but the expected statistics of the behavior in manufactured ICs.

Memory circuits tend to be the hardest hit by the problem of these random variations because of their high replication count on any single chip, which demands a very high statistical quality from the product. Requirements of 5–6σ (0.6 ppm to 2 ppb failure rate) are very common for today's SRAM caches of 1–10 Mb range, and these requirements are only further increasing with increasing amount of memory per die. Estimating such extreme statistics efficiently in simulation has been a challenging task until recent years. Many promising techniques have been proposed in recent years to simulate the statistical quality of memory designs. This book draws on the expertise of the pioneers in this topic of extreme statistical design and presents a comprehensive coverage of topics relevant to the simulation of the statistical quality of memory circuits. Topics covered include sources of variability in nanoscale semiconductor devices, algorithms for efficient estimation of extreme statistics and design approaches that exploit such statistical information.

The editors hope that this book will provide impetus to the practice and research of statistical design of VLSI electronics.

Yorktown Heights, NY   Amith Singhee
Urbana, IL   Rob A. Rutenbar

Contents

Contributors

Robert C. Aitken ARM Research, San Jose, CA, USA

Asen Asenov Department of Electronics and Electrical Engineering, The University of Glasgow, Lanarkshire G12 8QQ, UK

Wei Chen Northwestern University, Evanston, IL, USA

Xiaoping Du Missouri University of Science and Technology, Rolla, MO, USA

Chenjie Gu University of California, Berkeley, CA, USA

Rajiv Joshi IBM, Thomas J. Watson Research Center, Yorktown Heights, NY, USA

Rouwaida Kanj IBM Austin Research Labs, Austin, TX, USA

Jaijeet Roychowdhury University of California, Berkeley, CA, USA

Rob A. Rutenbar Dept. of Computer Science, University of Illinois at Urbana-Champaign, Urbana, IL, USA

Amith Singhee IBM, Thomas J. Watson Research Center, Yorktown Heights, NY, USA

Yu Wang Nanjing University of Aeronautics and Astronautics, Nanjing, China

Robert C. Wong IBM Microelectronics Division, Hopewell Junction, NY, USA

Chapter 1
Introduction

Amith Singhee

1.1 Yield-Driven Design: Need for Accurate Yield Estimation

Tradition VLSI circuit design would contend with three general objectives:

1. *Maximize performance*: In the context of an SRAM bitcell, this may be captured by several different metrics, for example, the read current supplied by the cell, or the read access time, given a specific peripheral circuit.
2. *Minimize power consumption*: For an SRAM bitcell, this could be, for instance, the active current drawn by the cell during a read operation, or the total leakage current drawn by the cell during standby.
3. *Maximize noise and noise susceptibility*: Several metrics may measure robustness of the circuit operation to noise. Again for an SRAM bitcell, a measure can be the static noise margin (SNM) of the cell.

These objectives are typically competing. For example, a higher read current (better performance) may require the cell transistors to have fast transistors, with low threshold voltages, but could in turn increase the leakage current (worse power consumption) and decrease immunity to noise (worse noise immunity). Any design exercise would be an attempt to balance optimally the circuit in terms of these three considerations, so as to satisfy the needs of the specific application.

The last decade has seen a large growth in the amount of manufacturing variations in VLSI circuits. As a result, good manufacturing yield is no longer guaranteed easily, and must be explicitly optimized for during the design phase. This gives us a fourth design objective:

4. *Maximize circuit yield*: The yield can be defect yield, resulting from hard defects like shorts/opens, or it can be parametric yield, resulting from variations in device characteristics, or some combination. In a well-patterned SRAM bitcell,

A. Singhee
IBM, Thomas J. Watson Research Center, Yorktown Heights, NY, USA
e-mail: asinghee@us.ibm.com

A. Singhee and R.A. Rutenbar (eds.), *Extreme Statistics in Nanoscale Memory Design*, Integrated Circuits and Systems, DOI 10.1007/978-1-4419-6606-3_1,

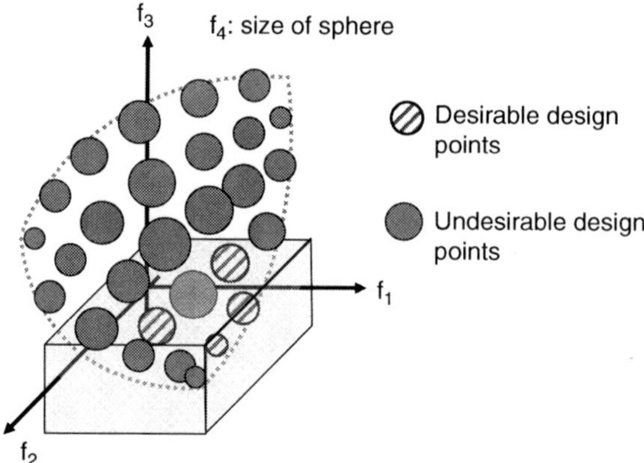

Fig. 1.1 Memory design must solve a four-dimensional tradeoff problem: (1) f_1: maximize performance, (2) f_2: minimize power, (3) f_3: maximize yield, and (4) f_4: maximize noise immunity. Larger positive values represent *worse* objective values in this figure: so we want to be as close to the origin as possible, with as small a sphere as possible. As a result, we may have one or more acceptable solutions in the space of solutions, given certain required thresholds for each f_i. The interior of the *box* defines acceptable values of (f_1, f_2, f_3), and acceptable f_4 values correspond to *sphere* smaller than some threshold

yield loss may be primarily determined by mismatch between the two halves of the cell due to random dopant fluctuation (RDF), for example.

Now we have a four-dimensional tradeoff in the design problem. Figure 1.1 shows an example of some design points in such a tradeoff. The goal of the design exercise is to find one or more of the desired design points, which meet some requirements for each of our objectives.

Traditional approaches of only implicitly considering yield by placing extra margins in the requirements for performance, power, and noise are not feasible anymore. This is because today's aggressive designs, especially highly replicated circuits like memory cells and latches, do not have large margins available for these objectives. Hence, the design exercise must explicitly consider yield as an objective or a constraint. A direct consequence of this yield-driven design methodology is that the designer requires simulation tools to *accurately* predict the parametric yield of the circuit, given a particular set of design parameters and variational parameters.

For transistor-level design, Monte Carlo simulation has been widely applied to predict the yield. The case for memory cells is, however, different, as we see next.

1.2 The Case of High-Replication Circuits

Much of the VLSI design effort across the industry is spent on designing memories. These include static random access memory (SRAM), dynamic RAM (DRAM), embedded DRAM, and non-volatile memories (NVM) like NOR flash. These memories are typically arrays of millions of memory cells. We refer to circuits, such as these memory cells, as High-Replication Circuits (HRCs) because they have the unique feature of being highly replicated in a single die. Other circuits like latches and flip-flops also fall in this class of HRCs, now that we may see millions of instances of a latch in an aggressive multi-core microprocessor.

An increasingly difficult aspect of designing these HRCs is meeting the extremely high parametric yields required for them. The high yield requirement is imposed by their high (and increasing) replication count. For example, IBM's Power6^{TM} processor, released in 2007, already uses 32 MB of L3 cache. To maintain a high functional yield of such a large array, the bitcells must have an extremely low failure probability – of the order of one failing cell per billion cells! This requirement is made difficult by the increasing manufacturing variation as semiconductor technology moves to deep sub-resolution feature sizes of 32 nm and smaller. For example, the threshold voltage variation at the 22-nm technology node is predicted to be 58% by the International Technology Roadmap for Semiconductors (ITRS 2007 [1]). These variations stem from several possible sources, such as RDF and line edge roughness (LER). The memory cells face the worst of this variation because they typically have the smallest devices and the tightest layouts in the entire chip. The widely used Pelgrom's law [2] relates the standard deviation of threshold voltage (V_t) to the device area as

$$\sigma(V_t) = \frac{k}{\sqrt{LW}} \tag{1.1}$$

where L and W are the length and width of the transistor, respectively. As device areas shrink with technology, the factor k does see some improvement because of process improvements. However, it is getting harder to keep scaling k with the feature size, resulting in the expected upward trend of $\sigma(V_t)$.

Meeting such extreme yield requirements with such high variability is a daunting challenge. A significant component of this challenge has been to estimate in simulation the failure probability (or, equivalently, the yield) of a memory cell design. Standard Monte Carlo fails miserably here because it requires extremely high computing resources to obtain estimates of these *extreme statistics*. This difficulty has led to a burst of exciting research and recent development of techniques to address this problem. This book attempts, for the first time, to provide a comprehensive and in-depth coverage of these techniques.

1.3 Why Does Standard Monte Carlo Not Work?

Monte Carlo simulation is the most general and accurate method for computing the failure probability (or yield) of any circuit and is widely used across the semiconductor industry. It does not, however, scale well to the problem of estimating extreme statistics. We briefly outline the algorithm and its pitfalls here. For a more detailed discussion of Monte Carlo, see [3] for instance.

1.3.1 Process Variation Statistics: Prerequisites for Statistical Analysis

To obtain any statistics of the circuit performance (e.g., yield), we first need to know the statistics of the relevant device or circuit parameters. For example, to estimate the statistics of the SNM of an SRAM cell using SPICE simulations, we need to know the statistics of, say, the threshold voltage variation for PMOS and NMOS transistors in the manufacturing process used, due to RDF. It is often observed that this variation follows a Gaussian distribution [4]. As another example, via or contact resistance may follow some non-Gaussian distribution [5]. We name all such parameters as *statistical parameters*. Some process characterization is required to estimate these probability distributions. The method of estimation may depend on how mature the technology is. In early stages of technology development, these distributions may be extrapolated from the previous technology, or from TCAD, as in [6] or [7]. In later stages, they may be characterized using hardware measurements, as in [4].

Let us say that we have s statistical parameters for our memory cell. For example, if we are considering only RDF in an 6T SRAM cell, we would have $s = 6$ threshold voltage deviations as the statistical parameters, each having a Gaussian distribution, as per [4]. We denote each parameter by x_i, where $i = 1,\ldots, s$. Together the entire set is denoted by the vector $\mathbf{x} = \{x_1,\ldots, x_s\}$. Each parameter has a probability distribution function (PDF) denoted by $\pi_i(x_i)$, and the joint PDF of $\mathbf{x}$ is denoted by $\pi(\mathbf{x})$.

1.3.2 Monte Carlo Simulation

Given certain a value for each of the s statistical parameter, we can enter these values into a SPICE netlist and run a SPICE simulation to compute the corresponding value of any circuit performance metric. For example, we can compute the SNM of a 6T SRAM cell given the six threshold voltage values. Let us say the performance metric is y and we have some specification on the performance metric; for example, the write time for a memory cell must be less than 50 ps. Let us denote this specification value by t, and the requirement by $y \leq t$. Note that here we use $\leq$ without any loss of generality. Now, for certain combinations of values for the statistical parameters, our circuit will fail this specification; that is, y will be greater than t. For example, if the threshold voltage is too large, the access time may exceed the maximum allowed value due to slower devices.

To estimate the yield of the circuit using Monte Carlo simulation, we *randomly* draw values for all the s statistical parameters from their joint PDF. One such set of *s* random values is called a *sample point*, or simply a *sample*. We generate several such samples until we obtain several failing points. Suppose that we generate n total points and obtain n_f failing points. The *estimate* of circuit failure probability is then given by

$$\hat{F}_f = \frac{n_f}{n} \qquad (1.2)$$

and the estimate of the circuit yield is $1 - \hat{F}_f$.

Note that if we re-run Monte Carlo for the same number of points, but with a different random number sequence, we will obtain a different estimate of F_f. Hence, there is some statistical error in the Monte Carlo estimate, and this error decreases with increasing number of points n [3]. Since we run a full SPICE simulation for each sample point, the cost of an n-point Monte Carlo run is roughly the same as that for n SPICE simulations, assuming that the cost of generating random numbers and using (1.2) is negligible in comparison.

1.3.3 The Problem with Memories

Consider the case of a 1 megabit (Mb) memory array, which has 2^{10} "identical" instances of a memory cell. To simplify our discussion, we will assume that there is no redundancy in the memory. These cell instances are designed to be identical, but due to manufacturing variations, they will differ. Suppose we desire a *chip* yield of 99%; that is, no more than one chip per 100 should fail. We will see in the next chapter that this translates to a maximum *cell* failure rate of 9.8 cells per billion cells.

If we want to estimate the yield of such a memory cell, a standard Monte Carlo approach would require at least 100 million SPICE simulations on average to obtain just one failing sample point! Even then, the estimate of the yield or failure probability will be suspect because of the lack of statistical confidence, the estimate being computed using only one failing example. Such a large simulation count is usually intractable. This example clearly illustrates the problem with designing robust memories in the presence of process variations: we need to simulate *rare* or *extreme* events and estimate the statistics of these rare events. The problem of simulating and modeling rare events stands for any high-replication circuit, like SRAM cells, eDRAM cells and non-volatile memory cells.

1.4 An Overview of This Book

This book attempts to provide a comprehensive treatment of recent research developments that are relevant to the general problem of estimating extreme statistics, with a strong focus on memory design. The purpose of this book is to consolidate existing ideas in this field of research to provide a clear reference for practitioners and students, and a solid footing for further research in improving memory design methodology. The topics covered in this book fall under the following general categories:

1. Techniques for estimating extreme statistics – Chapters 4, 5, 6, 7 and 8.
2. Parametric variation in transistors: modeling, simulation and impact on design – Chapter 3.
3. Memory design methodology: statistical and electrical metrics – Chapters 2, 5 and 7.

A chapter-wise outline of the book is as follows:

Chapter 2: Extreme Statistics in Memory Design
Author: Amith Singhee
This chapter provides a mathematical treatment of the statistics of memory arrays. It relates array-level yield requirements to the yield requirements of a single memory cell and shows that non-extreme array yield requirements result in extreme yield requirements for the memory cell. It also discusses the impact of redundancy in the memory on these cell-level yield requirements.

Chapter 3: Statistical Nano CMOS Variability and Its Impact on SRAM
Author: Asen Asenov
Yield estimating requires accurate modeling of device variability. Chapter 3 discusses these variations in detail. It describes the dominant sources of statistical variability in complementary MOS (CMOS) transistors at and below 45 nm. In particular, it discusses device degradation mechanisms, manufacturing imperfections like RDF and LER, and the interaction between these phenomena. It analyses the combined impact of these variation sources on CMOS, ultra thin body silicon-on-insulator (UTBSOI), and finFET devices in future technologies. It discusses compact modeling strategies that enable yield estimation using SPICE simulation and predicts the implications of these variations on SRAM design up to the 13-nm technology generation.

Chapter 4: Importance Sampling-Based Estimation: Applications to Memory Design
Authors: Rouwaida Kanj and Rajiv Joshi
This chapter provides an in-depth discussion of Monte Carlo-based estimation. It discusses different techniques for implementing sampling-based estimation methods and performing statistical inference. It describes the importance sampling technique as an efficient sampling-based method that can be used to estimate extreme statistics and discusses its application to the memory design problem.

Chapter 5: Direct SRAM Operation Margin Computation with Random Skews of Device Characteristics
Author: Robert C. Wong
This chapter draws on the extensive industrial design experience of the author to discuss practical static and dynamic metrics for qualifying SRAM bitcell designs and associated simulation techniques. It describes direct methods for computing

statistical yield margins and their application to a yield-driven SRAM design methodology.

Chapter 6: Yield Estimation by Computing Probabilistic Hypervolumes
Authors: Chenjie Gu and Jaijeet Roychowdhury
This chapter presents the YENSS method, which approaches the yield estimation problem as a volume integration problem. It employs line search techniques to identify points on the boundary between failed and successful operation of the circuit. It maintains a simplicial approximation of the successful operation region and iteratively refines it as more points are identified. The volume of this simplicial approximation then gives an easily computable estimate of the yield.

Chapter 7: Most Probable Point-Based Methods
Authors: Xiaoping Du, Wei Chen, and Yu Wang
Most probable point (MPP)-based methods for yield estimation have found wide application in reliability-driven design, largely outside the domain of VLSI circuit design. These methods are in principle equally applicable to memory design when appropriately adapted. This chapter discusses some of these methods, in particular, the first order reliability method (FORM), the second order reliability method (SORM) and, in brief, the first order saddlepoint approximation (FOSPA) method. The chapter uses worked examples to illustrate the FORM and SORM.

Table 1.1 Methods for estimating extreme memory statistics, covered in this book

Method	Characteristics
Importance sampling	• Circuit independent
	• Avoids assumptions
	• Sampling based
Direct SRAM margin computation	• Exploits circuit characteristics
	• Deterministic approach
	• Provides design insight
	• Specific parameters and metrics
YENSS	• Circuit independent
	• Deterministic
	• Good for small dimensionality
Most probable point-based methods	• Circuit independent
	• Simplifying assumptions
	• Deterministic
Memory margining with EVT	• Circuit independent
	• Sampling based
	• Simplifying assumptions
	• Provides design insight
	• Provides probability distribution
Statistical blockade	• Circuit independent
	• Avoids assumptions
	• Sampling based
	• Provides probability distribution

Chapter 8: Extreme Value Theory: Application to Memory Statistics
Authors: Robert C. Aiken, Amith Singhee, and Rob A. Rutenbar
This chapter presents relevant results from extreme value theory and nonparametric statistics and applies them to the problem of extreme statistics in memory design. It explains the problem of statistical margining for memories and presents two approaches: (1) memory margining using extreme value theory and (2) statistical blockade. The latter is a statistical simulation method that learns the failure region of the memory cell to reduce Monte Carlo simulation, and then employs extreme value theory to model tail statistics and estimate failure probabilities.

These chapters cover the whole gamut of approaches to the memory statistics estimation problem. Table 1.1 presents some distinguishing characteristics of each of these methods.

References

1. International Technology Roadmap for Semiconductors. http://www.itrs.net
2. Pelgrom MJM, Duinmaijer ACJ, Welbers APG (1989) Matching properties of MOS transistors. IEEE J Solid-State Circuits 24(5):1433–1440
3. Fishman GS (2006) A first course in Monte Carlo. Duxbury, Belmont
4. Agarwal K, Liu F, McDowell C, Nassif S, Nowka K, Palmer M, Acharyya D, Plusquellic J (2006) A test structure for characterizing local device mismatches. In: Symposium on VLSI circuits digest of technical papers, 2006
5. Chang H, Zolotov V, Narayan S, Visweswariah C (2005) Parameterized block-based statistical timing analysis with non-Gaussian parameters, nonlinear delay functions. In: Proceedings of IEEE/ACM design automation conference, 2005
6. Hane M, Ikezawa T, Ezaki T (2003) Atomistic 3D process/device simulation considering gate line-edge roughness and poly-Si random crystal orientation effects. In: Proceedings of IEEE international electron devices meeting, 2003
7. Ezaki T, Izekawa T, Hane M (2002) Investigation of random dopant fluctuation induced device characteristics variation for sub-100 nm CMOS by using atomistic 3D process/device simulator. In: Proceedings of IEEE international electron devices meeting, 2002

Chapter 2
Extreme Statistics in Memories

Amith Singhee

Memory design specifications typically include yield requirements, apart from performance and power requirements. These yield requirements are usually specified for the entire memory array at some supply voltage and temperature conditions. For example, the designer may be comfortable with an *array failure probability* of one in a thousand at $100°C$ and 1 V supply; i.e., $F_{f,array} \leq 10^{-3}$. However, how does this translate to a yield requirement for the *memory cell*? What is the maximum *cell failure probability*, $F_{f,cell}$, allowed so as to satisfy this array failure probability requirement? We will answer these questions and in the process understand the relevance of extreme statistics in memory design.

2.1 Cell Failure Probability: An Extreme Statistic

$F_{f,array}$, the chip or array failure probability, is the probability of *one or more* cells failing in any one array. Hence, if we manufacture ten arrays, and six of them have a total of eight faulty cells, our array failure probability is (approximately) 6/10, and not 8/10. Another equivalent way to define $F_{f,array}$ is: the probability of the *worst* cell in an array being faulty. Hence, if we are measuring some performance metric y (e.g., static noise margin), we are interested in the statistics of the worst value of y from among N values, where N is the number of cells in our array. Let us define this worst value as M_N, and let us assume that by worst, we mean the maximum; i.e., large values of y are bad. If small values of y are bad, then we can use the maximum of $-y$ as the worst value.

A. Singhee
IBM, Thomas J. Watson Research Center, Yorktown Heights, NY, USA
e-mail: asinghe@us.ibm.com

A. Singhee and R.A. Rutenbar (eds.), *Extreme Statistics in Nanoscale Memory Design*, Integrated Circuits and Systems, DOI 10.1007/978-1-4419-6606-3_2,

$$M_N = \max(Y_1, \ldots, Y_N),$$

where $Y_1, \ldots, Y_N$ are the measured performance metric values for the N cells in an array. Also suppose that the failure threshold for y is y_f; i.e., any cell with $y > y_f$ is defined as failing. Then, $F_{f,\text{array}}$ can be defined as

$$F_{f,\text{array}} = P(M_N > y_f), \tag{2.1}$$

that is, the probability of the worst case y in the array being greater than the failure threshold y_f. Now,

$$P(M_N > y_f) = 1 - P(M_N \leq y_f),$$

where $P(M_N \leq y_f)$ is the probability of the worst case cell passing: this is the same as the probability of all the cells in the array passing; that is,

$$P(M_N \leq y) = P(Y_1 \leq y_f, \ldots, Y_N \leq y_f). \tag{2.2}$$

Now, we assume that the failure of any cell in the array is independent of the failure of any other cell in the array. This is reasonable because we are modeling failures due to random variations, which largely tend to be spatially independent. For instance, a well-designed CMOS SRAM bitcell would fail from random variation mainly because of random mismatch between its left and right halves induced primarily because of random dopant fluctuation. It is well known that random dopant fluctuation in one transistor is largely independent of the fluctuation in another transistor.

Given this assumption of independence of cell failure events, we can write (2.2) as

$$P(M_N \leq y) = \Pi_{i=1}^{N} P(Y_i \leq y_f) = [P(Y \leq y_f)]^N. \tag{2.3}$$

Now,

$$P(Y \leq y_f) = 1 - P(Y > y_f), \tag{2.4}$$

where $P(Y_i > y_f)$ is the failure probability of a cell; i.e., $F_{f,\text{cell}}$. Hence, combining all equations from (2.1) to (2.4), we have

$$F_{f,\text{array}} = 1 - [1 - F_{f,\text{cell}}]^N \tag{2.5}$$

or

$$F_{f,\text{cell}} = 1 - [1 - F_{f,\text{array}}]^{\frac{1}{N}}. \tag{2.6}$$

2.1.1 Units of Failure Probability

There are some different units used to represent failure probability, especially in the context of extremely low probability. Some relevant ones are as follows:

- Raw probability (P): This is the most inconvenient representation, for example, A 32 Mb array with an array level requirement of $F_{f,array} < 0.1\%$ will need a cell failure probability $< 2.98 \times 10^{-11}$.
- Parts-per-million (ppm): This is the expected number of failure events (failing cells) in one million events (cells), and is given by $P \times 1$ million. So, a raw probability of 2.98×10^{-11} is the same as 2.98×10^{-5} ppm.
- Parts-per-billion (ppb): This is the expected number of failure events in one billion events and is given by $P \times 1$ billion. Our example of 2.98×10^{-11} would be 0.0298 ppb
- Equivalent sigma (σ): This is the $(1 - P)$th quantile on the standard normal distribution, and is given by

$$m = \Phi^{-1}(1 - P) = -\Phi^{-1}(P), \tag{2.7}$$

where Φ^{-1} is the inverse of the standard normal cumulative distribution function (CDF). So, if $P = 2.98 \times 10^{-11}$, we get $m = 6.5447\sigma$. This is the most widely used measure of extreme failure probability in the domain of memory design.

2.1.2 An Example of Extreme Statistics in Memories

Table 2.1 shows the cell failure probability in parts-per-billion, calculated for several values of array yield specification and several values of array size. Looking at the highlighted typical values, we can see that the cell yield requirements are very stringent: only a few cells per billion can fail. Apart from the significant difficulty in

Table 2.1 Cell failure probability $F_{f,cell}$, in parts-per-billion (ppb), computed from (2.6) for a range of array sizes N and array yield specifications $F_{f,array}$. Typical cases are highlighted in bold

N (Mb)	$F_{f,array}$ (%)			
	<10	<1	<0.1	<0.01
1 (2^{10})	102,886	9,815	977	97.7
2	51,444	4,907	489	48.8
4	25,722	2,454	**244**	**24.4**
8	12,861	1,227	**122**	**12.2**
16	6,431	613	**61.1**	**6.1**
32	3,215	307	**30.5**	**3.1**
64	1,608	153	**15.3**	**1.5**
128	804	76.7	7.6	0.76
256	402	38.3	3.8	0.38
512	201	19.2	1.9	0.19

design such a robust memory cell, it is difficult to even estimate in simulation, the fail probability of a given cell design with acceptable accuracy!

2.2 Incorporating Redundancy

A recent trend is to increase the fault tolerance of the memory by including redundancy in the array. A common approach is to have redundant columns in the array that are used to replace defective columns in the chip: see, for example, [1]. As array sizes increase, the chances of having a failing cell in one array increase. In reality, this failure can be due to radiation-induced soft errors, manufacturing defects, or process variation. Redundancy allows increased tolerance to such failures. In the simplest case, if there is one redundant column, the chip can tolerate one failing cell. The use of redundancy, of course, changes the relation between the array failure probability and the cell failure probability. We will develop a simple model for computing the required process-induced cell failure probability $F_{f,cell}$, given the maximum tolerable array failure probability $F_{f,array}$, when we have redundancy in the array.

Let us say that the amount of redundancy in the array is r. For the case of column redundancy, this would mean that we have r redundant columns, and can tolerate up to r faulty columns in the array. If each column has n_c cells, then we can tolerate up to rn_c faulty cells, depending on their spatial configuration. However, the probability of having two or more faulty cells in the same column can be shown to be negligibly small. Hence, we consider only the case where all the faulty cells in the array are in different columns. This will help us keep unnecessary and tedious mathematics away, without losing the core insights of the derivation. In this case, the array would operate without faults with up to r faulty cells in it, and the array failure probability is the probability of having more than r faulty cells in it. If the probability of having exactly k faulty cells in an array is denoted by $F_{f,k}$, then we can write the array failure probability $F_{f,array}$ as

$$F_{f,array} = 1 - (F_{f,0} + F_{f,1} + \cdots + F_{f,r}).$$
$$(2.8)$$

Now, all we need is to compute every $F_{f,k}$ in terms of the cell failure probability $F_{f,cell}$. We can write this $F_{f,k}$, the probability of k cells failing in an array, as

$$F_{f,k} = {}^{N}C_k F_{f,cell}^{k} (1 - F_{f,cell})^{N-k},$$
$$(2.9)$$

where ${}^{N}C_k$ is the number of combinations of k cells in an array of N cells. For $k = 0$, we get

$$F_{f,0} = (1 - F_{f,cell})^{N}.$$

Hence, if we have zero redundancy, we get from (2.8)

$$F_{\text{f,array}} = 1 - (1 - F_{\text{f,cell}})^N,$$

which is the same result as (2.5). As we would expect, we get the same result whether we use the worst case argument or zero redundancy argument, since we are modeling the same statistic. Using (2.9) in (2.8), we can obtain the array failure probability in the presence of redundancy. However, (2.9) is inconvenient because for N large (one million or more), the $^N C_k$ term can be inconveniently large, and the $F_{\text{f,cell}}^k$ term can be inconveniently small for floating point arithmetic.

2.2.1 The Poisson Yield Model

We can further simplify (2.8) to obtain a more popular model of the yield: the *Poisson model*. This model will avoid the unwieldy numerics of (2.9). Let us define λ as the expected number of faulty cells in every N cells we manufacture. Hence,

$$\lambda = F_{\text{f,cell}}N.$$

For example, if $F_{\text{f,cell}}$ is 1 ppm, and N is ten million, λ will be 1 ppm $\times$ 10 million $= 10$. Then, we can write (2.9) as

$$
\begin{aligned}
F_{\text{f},k} &= \frac{N!}{k!(N-k)!}\,\frac{\lambda^k}{N^k}\left(1 - \frac{\lambda}{N}\right)^{N-k} \\
&= \frac{N!}{(N-k)!\,N^k} \times \left(1 - \frac{\lambda}{N}\right)^N \times \left(1 - \frac{\lambda}{N}\right)^{-k} \times \frac{\lambda^k}{k!}.
\end{aligned}
\tag{2.10}
$$

There are now four terms that are multiplied together. Note that the array size N is very large compared to all relevant values of k and λ. For large values of N, we can approximate the first three terms in this equation as follows.

$$
\text{Term 1:}\left|
\begin{aligned}
\lim_{N\to\infty}\frac{N!}{(N-k)!\,N^k} &= \lim_{N\to\infty}\frac{N}{N}\,\frac{N-1}{N}\cdots\frac{N-k+1}{N} \\
&= \lim_{N\to\infty}1\left(1-\frac{1}{N}\right)\left(1-\frac{2}{N}\right)\cdots\left(1-\frac{k-1}{N}\right) = 1
\end{aligned}
\right.
$$

$$
\text{Term 2:}\left|
\begin{aligned}
\lim_{N\to\infty}\left(1-\frac{\lambda}{N}\right)^N &= \lim_{N\to\infty}1 + \sum_{i=1}^{\infty}\frac{N!}{i!(N-i)!}\frac{(-\lambda)^i}{N^i} \\
&= 1 + \sum_{i=1}^{\infty}\frac{(-\lambda)^i}{i!}\left(\lim_{N\to\infty}\frac{N!}{(N-i)!\,N^i}\right) = 1 + \sum_{i=1}^{\infty}\frac{(-\lambda)^i}{i!}(1) = e^{-\lambda}.
\end{aligned}
\right.
$$

Note here, that the limit expression of Term 1 emerges in the summation and we have replaced it with the limit value of 1.

$$\text{Term 3}: \left| \lim_{N \to \infty} \left(1 - \frac{\lambda}{N}\right)^{-k} = (1 - 0)^{-k} = 1.\right.$$

Using these limiting values for the first three terms in (2.10), we obtain

$$F_{f,k} \approx 1 \times e^{-\lambda} \times 1 \times \frac{\lambda^k}{k!} = \frac{\lambda^k e^{-\lambda}}{k!}.$$

Hence, we see that $F_{f,k}$ follows the well known *Poisson process*. We can finally write the array failure probability $F_{f,\text{array}}$ as

$$F_{f,\text{array}} \approx 1 - \sum_{k=0}^{r} \frac{\lambda^k e^{-\lambda}}{k!} = 1 - \sum_{k=0}^{r} \frac{(F_{f,\text{cell}}N)^k e^{-F_{f,\text{cell}}N}}{k!}. \tag{2.11}$$

If we have a specification for $F_{f,\text{array}}$, we can easily compute the specification for $F_{f,\text{cell}}$ numerically.

2.2.1.1 An Example: Quantifying Fault Tolerance with Statistical Analysis

Suppose that we require an array yield of at least 0.9; that is, $F_{f,\text{array}} = 0.1$ (or less), and our array has $N = 10$ million cells with additional redundant columns. Using (2.11), we can easily study the impact of redundancy on the required cell yield: we can compute the cell failure probability $F_{f,\text{cell}}$ for different values of r. The results are shown in Table 2.2. Clearly, increasing the redundancy increases the fault tolerance, and we can use cells with higher failure probability. However, the real measure of increase in fault tolerance is the percentage increase in cell failure probability allowed by an increase in redundancy. This is shown in column 3, and it is clear that adding redundancy helps, but with diminishing returns. At some point

Table 2.2 Cell failure probability $F_{f,\text{cell}}$, computed from (2.11) for a 10 Mb array and a cell failure probability $F_{f,\text{array}}$ of 0.1. As the fault tolerance (redundancy) increases, we can tolerate much larger cell failure probability, but with diminishing returns

Number of faults tolerable (r)	Maximum allowed $F_{f,\text{cell}}$ (ppm)	Percent increase in $F_{f,\text{cell}}$
0	0.0105	–
1	0.0532	404.8
2	0.1102	107.2
3	0.1745	58.32
4	0.2433	39.42
5	0.3152	29.57
6	0.3895	23.57
7	0.4656	19.55
8	0.5432	16.67

we reach an optimal trade-off point where further increase in fault tolerance is too small to justify the area, delay and power costs of adding redundant circuitry. Such critical, yield-aware analysis of design trade-offs is only possible with a rigorous statistical analysis methodology.

From the preceding discussion, we now know how to compute the specification on the cell failure probability, given a specification on the array failure probability. Suppose now that we design a cell. How do we estimate its failure probability to check if it meets this criterion? This will be the focus of the following chapters.

Reference

1. Joshi B, Anand RK, Berg C, Cruz-Rios J, Krishnamurthi A, Nettleton N, Nguyen S, Reaves J, Reed J, Rogers A, Rusu S, Tucker C, Wang C, Wong M, Yee D, Chang J-H (1992) A BiCMOS 50 MHz cache controller for a superscalar microprocessor. In: International Solid-State Circuits Conference, 1992

Chapter 3
Statistical Nano CMOS Variability and Its Impact on SRAM

Asen Asenov

3.1 Introduction

The years of "happy scaling" are over and the fundamental challenges that the semiconductor industry faces at technology and device level will deeply affect the design of the next generations of integrated circuits and systems. The progressive scaling of CMOS transistors to achieve faster devices and higher circuit density has fueled the phenomenal success of the semiconductor industry – captured by Moore's famous law [1]. Silicon technology has entered the nano CMOS era with 35-nm MOSFETs in mass production in the 45-nm technology generation. However, it is widely recognised that the increasing variability in the device characteristics is among the major challenges to scaling and integration for the present and next generation of nano CMOS transistors and circuits. Variability of transistor characteristics has become a major concern associated with CMOS transistors scaling and integration [2, 3]. It already critically affects SRAM scaling [4], and introduces leakage and timing issues in digital logic circuits [5]. The variability is the main factor restricting the scaling of the supply voltage, which for the last four technology generations has remained virtually constant, adding to the looming power crisis.

This chapter will focus on statistical variability, which has become a dominant source of variability for the 45-nm technology generation and which cannot be reduced by tightening process control. While in the case of systematic variability the impact of lithography and stress on the characteristics of an individual transistor can be modelled or characterised and therefore factored into the design process, in the case of statistical variability, only the statistical behaviour of the transistors can

A. Asenov
Department of Electronics and Electrical Engineering, The University of Glasgow, Glasgow, Lanarkshire G12 8QQ, UK
e-mail: A.Asenov@elec.gla.ac.uk

A. Singhee and R.A. Rutenbar (eds.), *Extreme Statistics in Nanoscale Memory Design*, Integrated Circuits and Systems, DOI 10.1007/978-1-4419-6606-3_3,

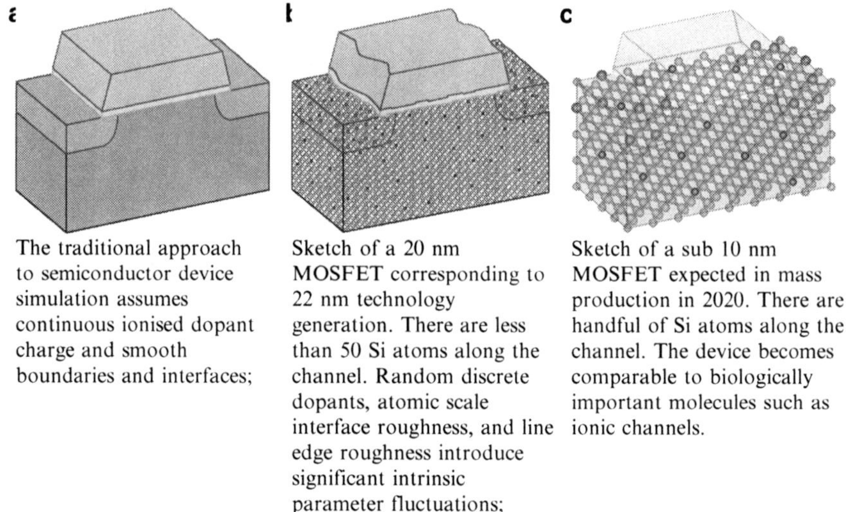

The traditional approach to semiconductor device simulation assumes continuous ionised dopant charge and smooth boundaries and interfaces;	Sketch of a 20 nm MOSFET corresponding to 22 nm technology generation. There are less than 50 Si atoms along the channel. Random discrete dopants, atomic scale interface roughness, and line edge roughness introduce significant intrinsic parameter fluctuations;	Sketch of a sub 10 nm MOSFET expected in mass production in 2020. There are handful of Si atoms along the channel. The device becomes comparable to biologically important molecules such as ionic channels.

Fig. 3.1 Transition from continuous towards "atomistic" device concepts

be simulated or characterised. Two adjacent macroscopically identical transistors can have characteristics from the two distant ends of the statistical distribution.

Figure 3.1 shows that MOSFETs are becoming truly atomistic devices. The conventional way of describing, designing, modelling and simulating such a semiconductor device, illustrated in Fig. 3.1a, assuming continuous ionised dopant charge and smooth boundaries and interfaces, is no longer valid. The granularity of the electric charge and the atomicity of matter, as illustrated in Fig. 3.1b, begin to introduce substantial variation in individual device characteristics. The variation in the number and position of dopant atoms in the active region of decanano MOS-FETs makes each transistor microscopically different and already introduces significant variations from device to device. In addition, the gate oxide thickness becomes equivalent to several atomic layers, with a typical interface roughness of the order of 1–2 atomic layers. This will introduce a localised variation in the oxide thickness, resulting in each transistor having a microscopically different oxide thickness or body thickness pattern. The granularity of the photoresist, together with other factors, will introduce unavoidable line edge roughness (LER) in the gate pattern definition and statistical variations in geometry between devices. The granularity of the polysilicon or the metal gate and the granularity of the high-k dielectric introduced in the 45-nm technology generation are other prominent sources of statistical variability.

In this chapter, we examine in detail how this statistical variability influences the device technology and scaling and how this process is reflected in the most recent update of the International Technology Roadmap for Semiconductors (ITRS). In the next section, we provide a classification of the different sources of statistical variability in contemporary MOSFETs. In Sect. 3.3, we review the major sources of statistical variability in nano CMOS transistors focusing at the 45-nm technology generation and beyond. Section 3.4 is dedicated to statistical reliability which acts in concert with the statistical reliability sources, resulting in rare but dramatic changes in the device characteristics in the process of degradation. The simulation techniques that are used to forecast the statistical variability in the future technology generations, including Drift-Diffusion, Monte Carlo and Quantum transport approaches, are reviewed in Sect. 3.5. Using the simulation techniques described in Sect. 3.5, Sect. 3.6 forecasts the magnitude of statistical variability in contemporary and future CMOS devices with conventional (bulk) and novel device architecture including ultra thin body (UTB) silicon on insulator (SOI) MOSFETs and FinFETs. The compact model strategies suitable for capturing the statistical variability in industrial strength compact models such as BSIM and PSP are outlined in Sect. 3.7. Section 3.8 presents an example of statistical circuit simulation employing the statistical compact model strategies discussed in Sect. 3.7. Finally, the conclusions are drawn in Sect. 3.9.

3.2 Process Variability Classification

Often the term "process variability" is used to encapsulate the increasing CMOS variability phenomena. This is not accurate and often confuses designers even in the big semiconductor companies that have technology and design under a single roof. The terminology is even more confusing for designers from the chipless and fabless companies, which normally are more than an arm's length away from the technology development. Different components of the contemporary CMOS variability are listed in Table 3.1.

Table 3.1 Different components of CMOS variability

		Process	Environment	Temporal
Global		$<L_o>$ and $<W>$ $<$Layer thickness$>$ $\langle R \rangle$'s $<$Doping$>$ $<V_{body}>$	T environment range V_{dd} range	$<$NBTI$>$ Hot electron shifts
Local	Systematic	OPC	Self-heating	Distribution of NBTI
		Phase shift Layout-mediated strain Well proximity	IR drops	Voltage noise SOI V_{body} history Oxide breakdown history
	Statistical	Random dopants Line edge roughness Poly-Si granularity Interface roughness High k morphology		
Across-chip		Line width due to pattern density effects	Thermal hot spots due to non uniform power dissipation	Computational load-dependent hot spots

The old fashioned "slow" or "global" process variability has been with the semiconductor industry forever. It is related to inaccuracy in the process parameters control and nonuniformity of equipment, and results in slow variation of device dimensions, layer thicknesses and doping concentrations (and the corresponding electrical parameters) across the wafer and from wafer to wafer. Usually when a new technology is ramped up, the "global" process variability is high but due to the heroic efforts of the technology teams, it is significantly reduced with the technology maturing, and the corresponding yield is steadily increased.

The systematic layout-related variability grew up in importance somewhere around the 90-nm technology generations. It is partially related to the fact that 193-nm lithography is still miraculously used to print sub 50-nm features today. Even with optical proximity (OPC) and phase shift corrections, [6] this results, as illustrated in Fig. 3.2, in deviation of the real device geometry from the ideal rectangular shapes that the designers have in their minds when laying out their transistors. The shapes of supposedly identical transistors appear to be different after fabrication and the level of differences is determined by the OPC strategy and the physical layout environment. The introduction of strain [7] to enhance the device performance at the 90-nm technology generation made things worse from the variability point of view. The performance of devices with identical geometry now varies not only because of the lithography-related geometry variations but also because of layout determined strain variation due to different spacing between the devices, different distances to the shallow trench isolation and different numbers and position of contacts. There is, however, hope with the systematic variability due to the fact that it is predictable. Both the shape of the individual devices after the OPC and the impact of layout-induced strain variations can be simulated and accurately predicted. Once their impact on device performance is evaluated, this can be factored in the design process and the deterministic variability disappears. Indeed, this is in the heart of the design for manufacturability strategies and tools widely adopted by integrated device manufacturers (IDMs) and foundries [8]. Also, with the introduction of regularized design and dummy features, the impact of the systematic (predictable) variability is greatly reduced.

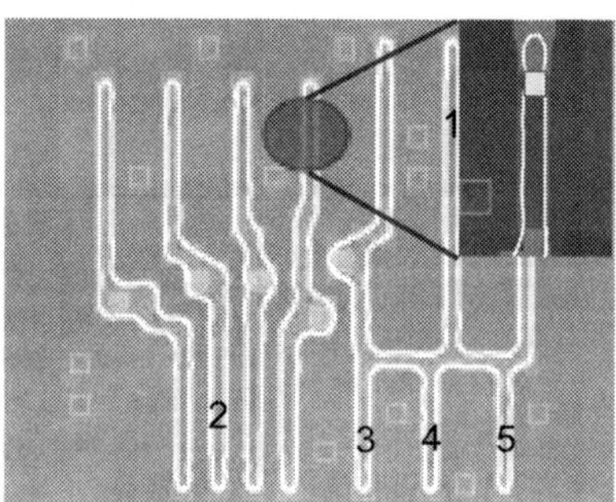

Fig. 3.2 Layout-induced variability

However, the most worrying of all is the rapidly increasing statistical variability introduced by discreteness of charge and granularity of matter in the transistors with decananometer dimensions. This is discussed in detail in the next section.

3.3   Sources of Statistical Variability

The statistical variability in modern CMOS transistors is introduced by the inevitable discreteness of charge and matter, the atomic scale non-uniformity of the interfaces and the granularity of the materials used in the fabrication of integrated circuits. The granularity introduces significant variability when the characteristic size of the grains and irregularities become comparable to the transistor dimensions. For conventional bulk MOSFETs, which are still the workhorse of the CMOS technology, random discrete dopants (RDD) are the main source of statistical variability [9]. Random dopants are introduced predominantly by ion implantation and redistributed during high temperature annealing. Figure 3.3 illustrates the dopant distribution obtained by the atomistic process simulator DADOS by Synopsys. Apart from special correlation in the dopant distribution imposed by the silicon crystal lattice, there may also be correlations introduced by the Coulomb interactions during the diffusion process. Line edge roughness (LER) illustrated in Fig. 3.4 stems from the molecular structure of the photoresist and the corpuscular nature of light. The polymer chemistry of the 193-nm lithography used now for few technology generations mainly determines the current LER limit of approximately 5 nm [10]. In transistors with polysilicon gate, poly gate granularity (PGG) illustrated in Fig. 3.5 is another important source of variability. This is associated surface potential pinning at the grain boundaries complimented by doping non-uniformity due to rapid diffusion along the grain boundaries [3].

The introduction of high-k/metal gate technology improves the RDD-induced variability, which is inversely proportional to the equivalent oxide thickness (EOT). This is due to the elimination of the polysilicon depletion region and better screening of the RDD-induced potential fluctuations in the channel from the very high concentration of mobile carriers in the gate. The metal gate also eliminates the

Fig. 3.3 KMC simulation of
RDD (DADOS, Synopsys)

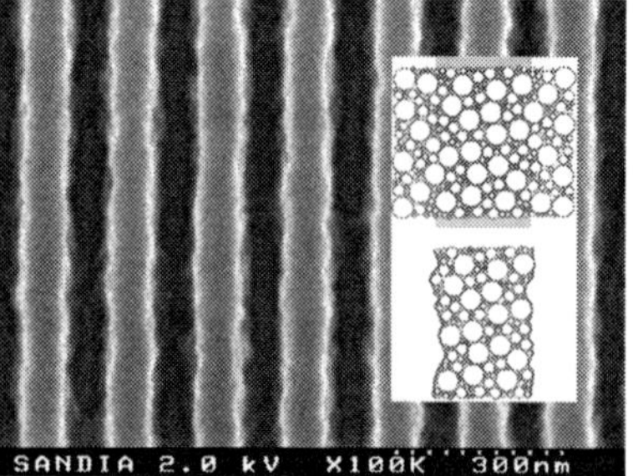

Fig. 3.4 Typical LER in photoresist (Sandia Labs)

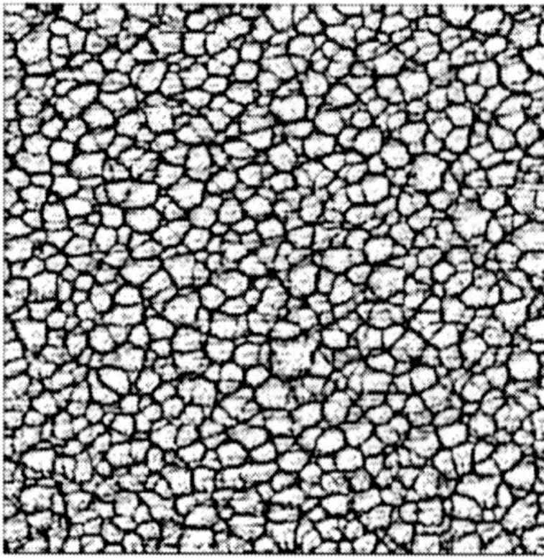

Fig. 3.5 SEM micrograph of typical PSG from bottom

PGG-induced variability. At the same time, it introduces high-k granularity illustrated in Fig. 3.6 and variability due to work-function variation associated with the metal gate granularity illustrated in Fig. 3.7 [11]. In extremely scaled transistors, atomic scale channel interface roughness illustrated in Fig. 3.8 [12] and corresponding oxide thickness and body thickness variations [13] can become important sources of statistical variability.

3.4 Statistical Aspects of Reliability

On top of statistical variability, problems related to statistical aspects of reliability are looming that will reduce the life-span of contemporary circuits from tens of years to 1–2 years, or less in the near future. In combination with RDD, which are the dominant source of statistical variability, the statistical nature of discrete trapped charges on defect states at the interface or in the gate oxide associated with hot electron degradation and negative/positive bias temperature instability (NBTI/PBTI) and hot carrier injection (HCI) result in relatively rare but

Fig. 3.6 Granularity in
HfON high-k dielectrics
(Sematech)

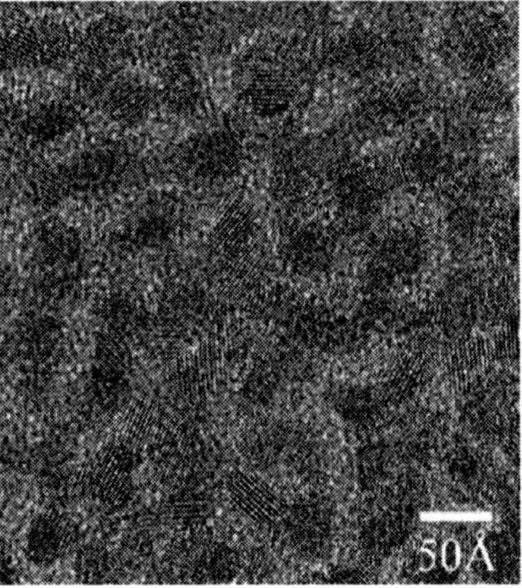

Fig. 3.7 Metal granularity
causing gate work-function
variation

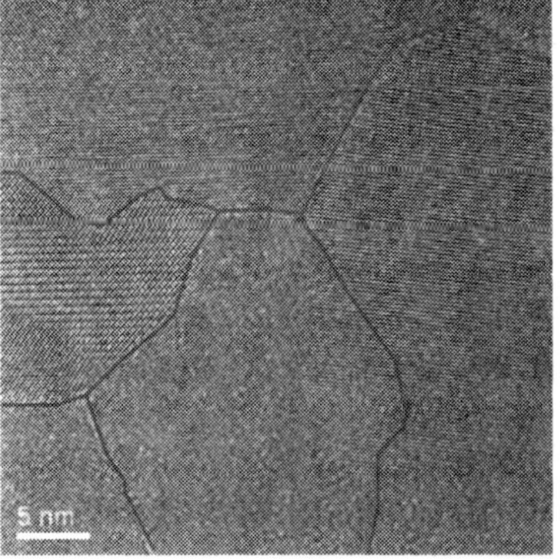

Fig. 3.8 Interface roughness
(IBM)

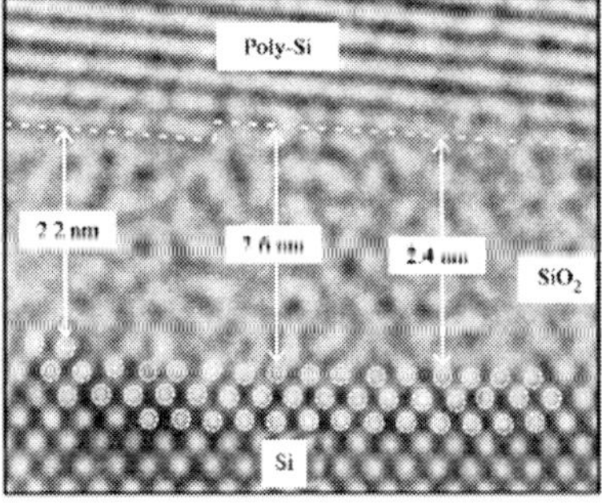

anomalously large transistor parameter changes, leading to loss of performance or
circuit failure [14]. For example, NBTI is one of the degradation mechanisms that
generate traps at the Si-channel/gate-oxide interface of MOSFETs during transistor

operation resulting in charge trapping, transistor parameters drift and circuit failures [15]. NBTI is associated with the Si–H bond breaking at the Si/SiO$_2$ interface and is triggered in negatively biased ($V_G < 0$ V, $V_S = V_D = 0$ V) p-channel MOSFETs at elevated temperatures present either due to power dissipation from the circuits or due to ambient temperature that the IC operates in. Carriers injected from the channel are captured by the traps. In large devices with many traps, this results in a well-defined change in the threshold voltage ΔV_T, which is determined by the equation

$$\Delta V_T = Q_T / C_{ox}, \tag{3.1}$$

where Q_T is the aerial density of the trapped charge and C_{ox} is the EOT. For NBTI which is the dominant degradation process in contemporary bulk p-channel MOSFETs with a conventional gate oxide, ΔV_T usually is in the tens of mV range.

In small devices, however, both the number of trapped charges and their positions vary from device to device. For example, in a 35-nm p-channel SRAM transistor corresponding to the 45-nm technology generation, a 1×10^{11} cm^2 NBTI trapped hole density results in an average of two trapped holes per transistor. However, the number of trapped holes follows a Poisson distribution and in reality varies from 0 to 10.

Figure 3.9 illustrates the RDD in a typical contemporary MOSFET with 35-nm channel length. An electron equiconcentration contour indicates the existence of a percolation path that carries most of the transistor current at threshold voltage. One or few strategically placed degradation-related charges could cut this current completely, introducing dramatic changes in the transistor parameters.

Figure 3.10 illustrates the statistical distribution of threshold voltage change due to single trapped electron or a Poissonian distribution with average, one which corresponds to an interface trapped charge density of 1×10^{11} cm^2 in the simulated square transistor. This corresponds to NBTI degradation that has created 1×10^{11} defects states per cm^2 at the Si/SiO$_2$ interface, all of which have trapped carriers injected from the channel of the transistor. Although the corresponding change in the threshold voltage calculated using (3.1) is only 5 mV, threshold voltage

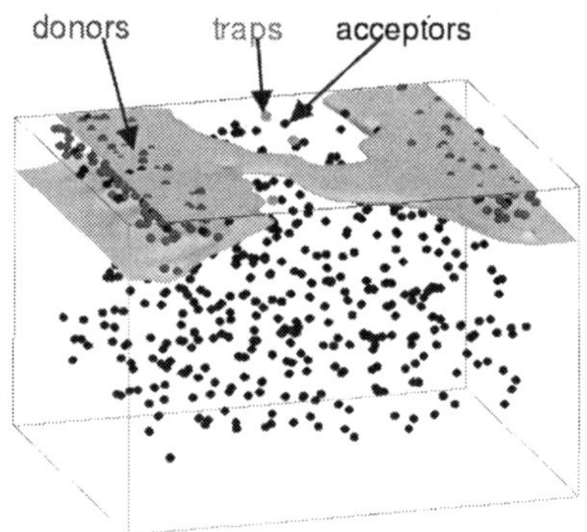

Fig. 3.9 Electron concentration distribution in a 30-nm "atomistic" MOSFET at threshold. Random donors, acceptors and traps marked

Fig. 3.10 Distribution of
threshold voltage change in a
35-nm MOSFET due to single
or multiple trapped electrons
corresponding to 1×10^{11} cm^2
trapped charge density

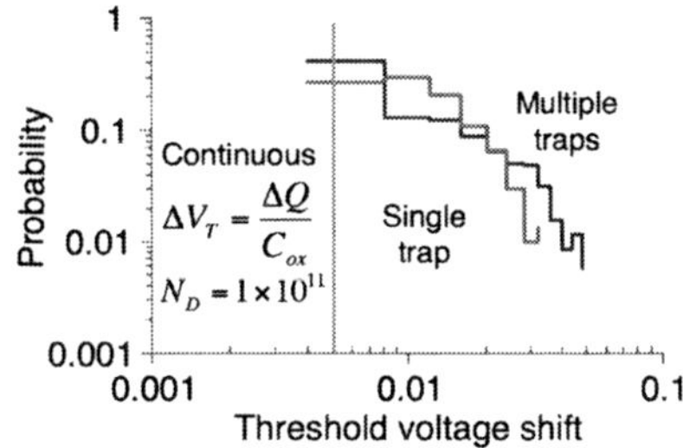

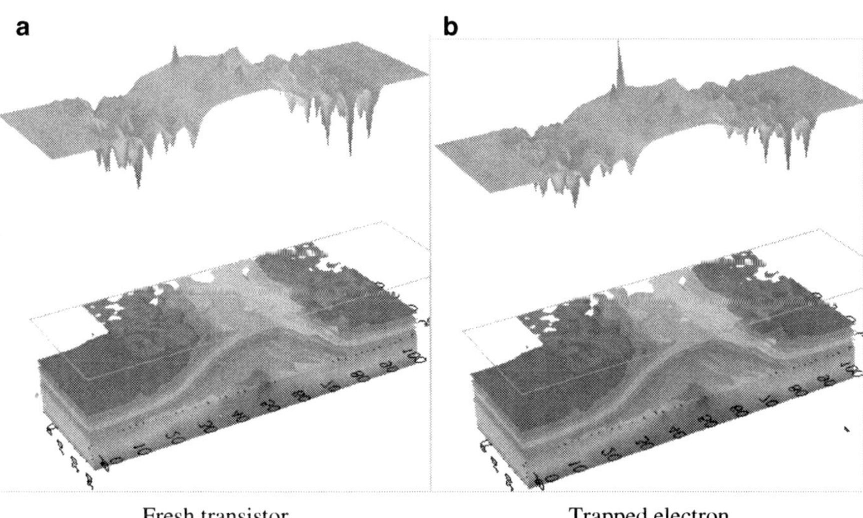

Fig. 3.11 Potential (*top*) and the electron concentration distribution (*bottom*) in a 35-nm (**a**) fresh
and (**b**) degraded transistor

changes larger than 30 mV are observed for a single trapped electron and larger
than 50 mV for a Poissonian distribution of trapped electrons. Figure 3.11 shows
the impact of a strategically positioned trap in the vicinity of a current percolation
path. Such percolation path is clearly visible in the upper part of the channel in
Fig. 3.11a. A single trapped charge, which produces the sharp pick in the potential
distribution in Fig. 3.11b, completely cuts the percolation path resulting in 30-mV
threshold voltage shift, which is six times larger than the average 5-mV threshold
voltage shift.

NBTI/PBTI and hot carrier degradation are already a fundamental problem in
flash [16] and SRAM memories [17], and starts to reduce the lifetime of digital
chips dramatically. This problem has been exacerbated by the introduction of high-k
gate stacks in mass production at the 45-nm technology generation in which case
PBTI becomes equally important to NBTI.

3.5   Gate Leakage Variability

In addition to the variability in the transistor characteristics, the discreteness of dopants in combination with atomic scale interface roughness introduce significant statistical variability in the gate leakage current as illustrated in Fig. 3.12 [18]. As illustrated in Fig. 3.13, 3D numerical simulations can trace the origin of these effects to local variations in the oxide thicknesses and electric field. For example, atomic scale random local reduction in the oxide thickness in the middle of the channel, seen in the interface roughness pattern in the middle contour in Fig. 3.13, results in significant increase in the local tunneling current density trough the gate oxide plotted in the top contour. The tunneling current is also locally mediated by the sharp increase in the electric field around the individual discrete dopants near the interface, illustrated at the bottom of Fig. 3.13.

In addition, the time-dependent generation of defect states at different positions of the complex structure of the gate oxide or the modern high-k dielectrics can result statistically in sharp localized increase in the leakage current openly

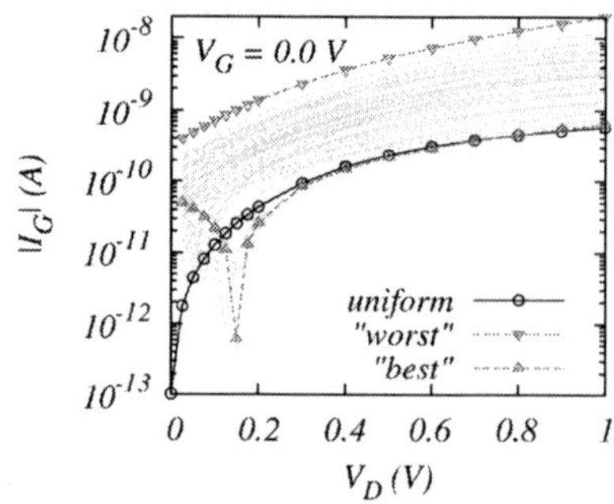

Fig. 3.12 Gate current variability in a 25-nm nMOSFET under the influence of random discrete dopants and atomic scale interface roughness

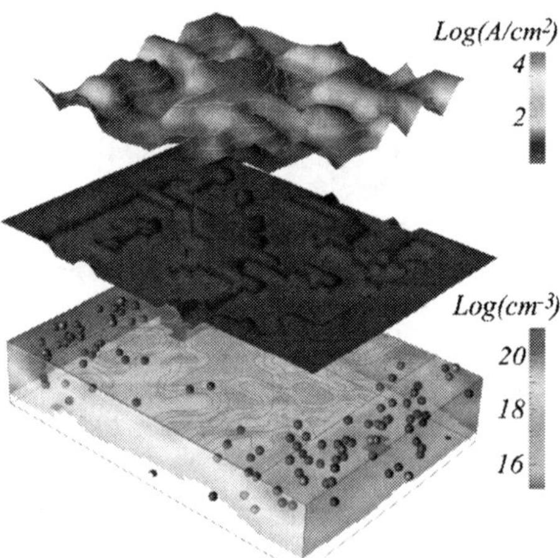

Fig. 3.13 Corresponding random dopant and potential distribution (*bottom*); interface roughness pattern (*middle*) and gate leakage current density (*top*)

interpreted as soft gate breakdown [19]. This is due to trap assisted tunneling through the NBTI, PBTI or HCI generated traps. The accumulation of these traps with time can also result randomly in irreversible hard oxide breakdown.

3.6   Simulation of Statistical Variability

The simulation results presented in this chapter were obtained using the Glasgow statistical 3D device simulator, which solves the carrier transport equations in the drift-diffusion approximation with Density Gradient (DG) quantum corrections [20]. In the simulations, the RDD are generated from continuous doping profile by placing dopant atoms on silicon lattice sites within the device S/D and channel regions with a probability determined by the local ratio between dopant and silicon atom concentration. Since the basis of the silicon lattice is 0.543 nm, a fine mesh of 0.5 nm is used to ensure a high resolution of dopant atoms. However, without considering quantum mechanical confinement in the potential well, in classical simulation, such fine mesh leads to carrier trapping at the sharply resolved Coulomb potential wells generated by the ionised discrete random dopants. In order to remove this artifact, the DG approach is employed as a quantum correction technology for both electrons and holes [14].

The LER illustrated in Fig. 3.4 is introduced through 1D Fourier synthesis. Random gate edges are generated from a power spectrum corresponding to a Gaussian autocorrelation function [10], with typical correlation length $\Lambda = 30$ nm and root-mean-square amplitude $\Delta = 1.3$ nm, which is the level that is achieved with current lithography systems [21]. The quoted literature values of LER are equal to 3Δ. The procedure used for simulating PGG involves the random generation of poly-grains for the whole gate region [3]: a large atomic force microscope image of polycrystalline silicon grains illustrated at the top of Fig. 3.5 has been used as a template and the image is scaled according to the average grain diameter experimentally observed through X-ray diffraction measurements made both in θ–2θ and θ scan modes [22] (the average grain diameter is 65 nm). Then the simulator imports a random section of the grain template image that corresponds to the gate dimension of the simulated device, and along grain boundaries, the applied gate potential in the polysilicon is modified in a way that the Fermi level remains pinned at a certain position in the silicon bandgap. In the worst case scenario, the Fermi level is pinned in the middle of the silicon gap. The impact of polysilicon grain boundary variation on device characteristics is simulated through the pinning of the potential in the polysilicon gate along the grain boundaries. The individual impact of RDD, LER and PSG on the potential distribution in a typical 35-nm bulk MOSFET is illustrated in Figs. 3.14–3.16, respectively.

The validation of our simulation technology is done in comparison with measured statistical variability data in 45-nm low power CMOS transistors [23]. The simulator was adjusted to match accurately the carefully calibrated TCAD device simulation results of devices without variability by adjusting the effective

Fig. 3.14 Potential distribution in a 35-nm MOSFET subject to RDD

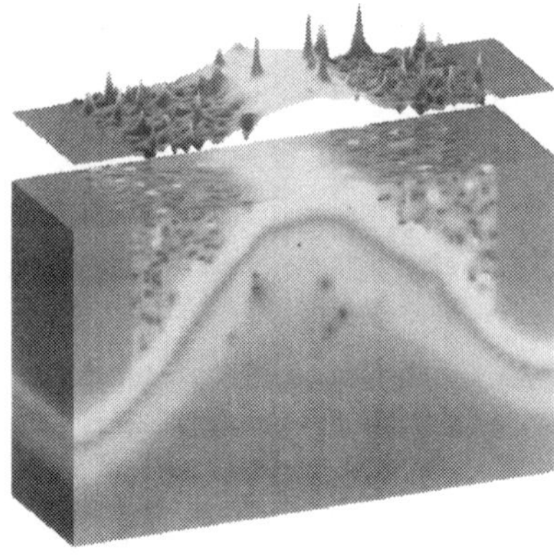

Fig. 3.15 Potential distribution in a 35-nm MOSFET subject to LER

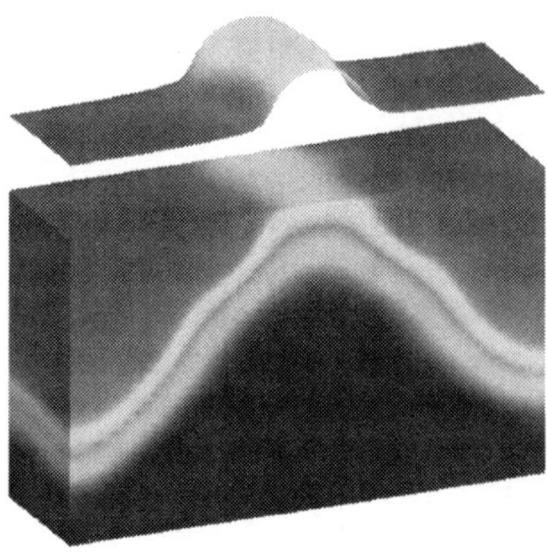

Fig. 3.16 Potential distribution in a 35-nm MOSFET subject to PSG

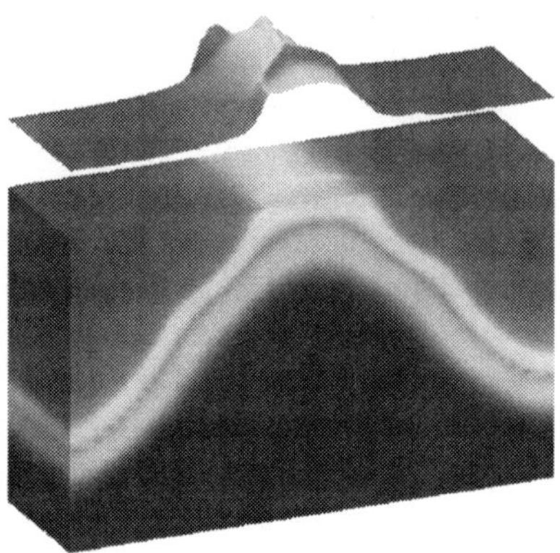

mass parameters involved in DG formalism, and the mobility model parameters. The calibration results are shown in Fig. 3.17, where "low" and "high" drain bias stand for 50 mV and 1.1 V, respectively.

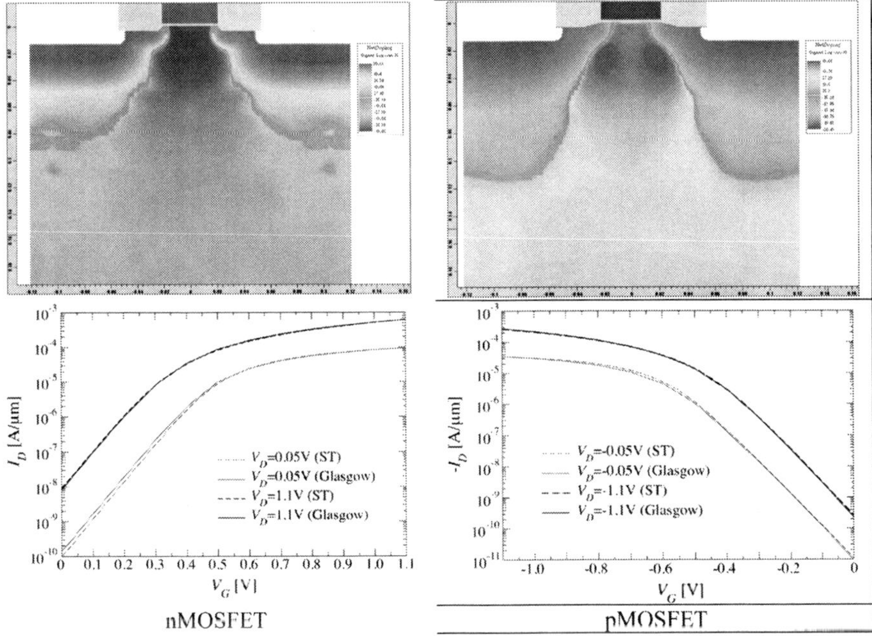

Fig. 3.17 *Top*: structure of the simulated 45-nm LP technology n(*left*) and p(*right*) channel MOSFETs; *Bottom*: Agreement between the commercial TCAD and the Glasgow "atomistic" simulator results

The potential distributions in the n- and p-channel transistors in Fig. 3.17, at gate voltage equal to the threshold voltage and low drain voltage of 50 mV, are illustrated in Fig. 3.18. In the n-channel transistor, RDD, LER and PSG are considered simultaneously, while in the p-channel transistor, only RDD and LER are considered. The electron concentration at the interface of the two transistors is mapped on top of the potential distributions. For example, acceptors in the channel of the n-channel transistor create sharp localized potential barrier for the electrons near the interface but act as potential wells for the holes in the substrate. At the same time, the donors in the source/drain regions create sharp potential well for electrons.

The simulation results for the standard deviation of the threshold voltage σV_T introduced by individual and combined sources of statistical variability are compared with the measured data in Table 3.2. In the *n*-channel MOSFET case, the accurate reproduction of the experimental measurements necessitates the assumption that in addition to RDD and LER, the PSG-related variability has to be taken into account. Constant current criterion in the subthreshold is used, defining the threshold voltage as the gate voltage producing drain current of $(I = 70$ nA.$W/L)$ at both low drain voltage of $V_{DS} = 0.05$ V and high drain voltage of $V_{DS} = 1.1$ V. Good agreement has been obtained by assuming that the Fermi level at the *n*-type poly-Si gate grain boundaries is pinned in the upper half of the bandgap at approximately 0.35 eV below the conduction band of silicon. However, in the

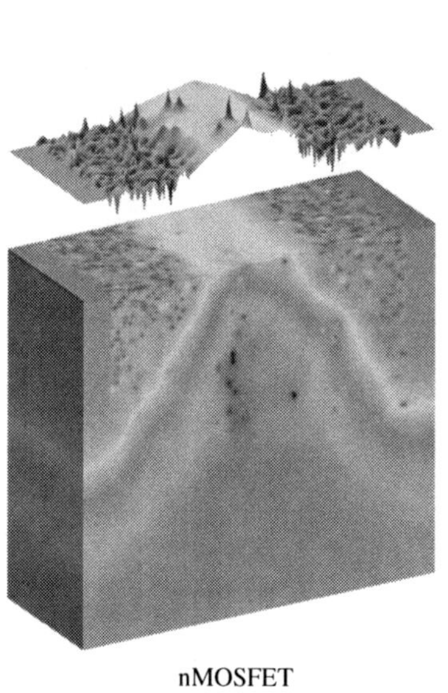
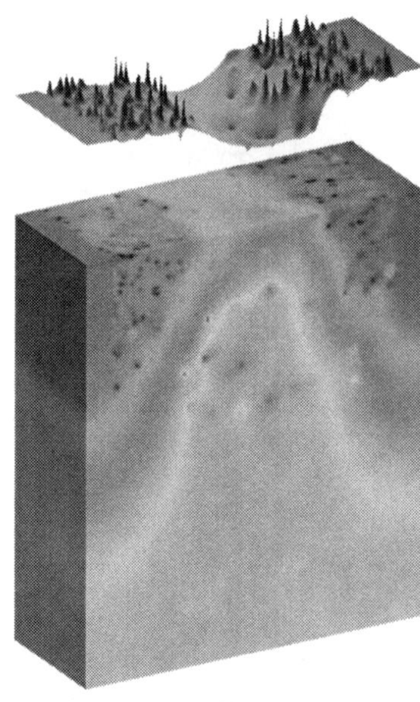

Fig. 3.18 *Top*: electron (*left*) and hole (*right*) concentration distribution at the interface; *Bottom*: potential distribution

Table 3.2 σV_T introduced by individual and combined sources of statistical variability

	n-channel MOSFET		*p*-channel MOSFET	
	σV_T [mV] ($V_{DS} = 0.05$ V)	σV_T [mV] ($V_{DS} = 1.1$ V)	σV_T [mV] ($V_{DS} = 0.05$ V)	σV_T [mV] ($V_{DS} = 1.1$ V)
RDD	50	52	51	54
LER	20	33	13	22
PSG	30	26	–	–
Combined	62	69	53	59
Experimental	62	67	54	57

p-channel MOSFET case, the combined effect of just the RDD and LER is sufficient to reproduce accurately the experimental measurements. The reason for this is the presence of acceptor-type interface states in the upper half of the bandgap which pins the Fermi level in the case of n-type poly-Si, and the absence of corresponding donor-type interface states in the lower part of the bandgap which leaves the Fermi level unpinned in the case of p-type poly-Si [24].

3.7 Simulation of Statistical Reliability

In the statistical NBTI simulations, the areal charge density associated with the threshold voltage shift observed in NBTI measurements of large self-averaging structures is translated into discrete trapped charges in microscopically different individual devices. Assuming that all the charges are trapped at the Si/SiO$_2$ interface, a fine, auxiliary 2D mesh is imposed at the interface. A rejection technique is used at each node of this mesh to determine if a single positive charge is located at that node or not based on the given interface charge sheet density. If it is determined that a single charge should be placed there, then its electronic charge is assigned to the surrounding nodes of the 3D discretisation mesh using a cloud-in-cell charge assignment scheme [25].

Figure 3.19 illustrates the results of the simulation of one particular microscopically different pMOSFET that includes random dopants and LER (a) without and (b) with additional trapped charges. Both the grey-shade mapped hole concentration and the positions of the random dopants are shown (acceptors in light grey, donors in dark grey). One iso-concentration surface is included providing a clear visual impression of the nonuniform channel formation due to the discrete random dopants and LER. In Fig. 3.19a, the channel initially forms on the right-hand-side (RHS) of the device, and the majority of the current near threshold flows through this region. This is due to the relatively small number of random dopants in this region. The surface plot above the device shows the hole concentration in a slice through the channel, indicating that the impact of each discrete dopant is accurately resolved. In Fig. 3.19b, a random set of interface trapped charges, with a sheet density of 5×10^{11} cm^{-2}, is added to the device of Fig. 3.19a. This random placement has resulted in two traps occurring at the RHS of the channel where

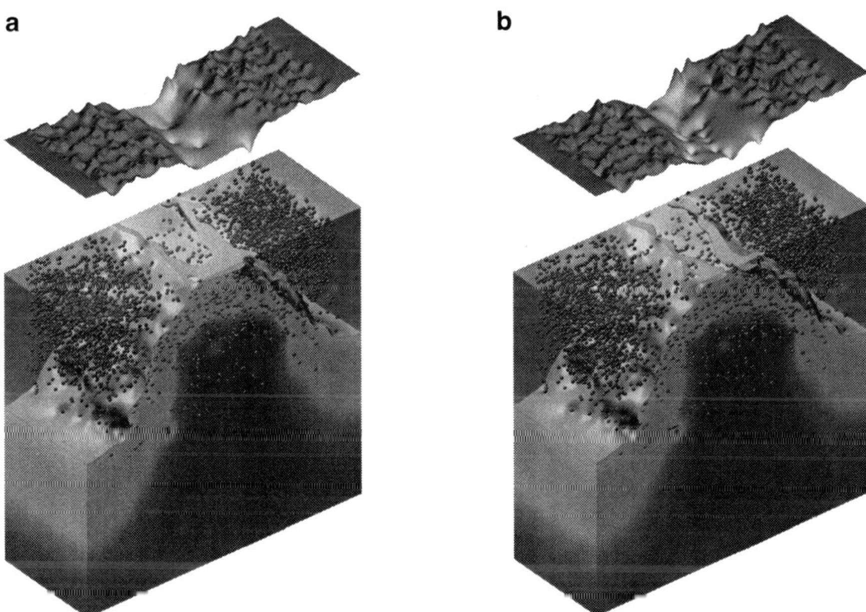

Fig 3.19 Hole concentration in one particular microscopically different p-channel device that includes random dopants and LER (a) without and (b) with additional trapped charges. The trapped charge sheet density is 5×10^{11} cm^{-2}. The extra trapped charges are visible in (b). One iso-concentration surface is included, and the surface plot above shows the hole concentration in a sheet through the channel

the current flow was predominant before. The charges trapped here effectively block this current path, as can be seen from the iso-concentration surface in this case. The result of this particular configuration of discrete dopants and traps is "anomalously" a large 140-mV increase in threshold voltage, when the expected "average" increase of the threshold voltage is approximately 40 mV.

The simulations are carried out on statistical samples of 200 low-power 45-nm devices. The fresh devices include RDD and LER. Three different levels of NBTI degradation are considered corresponding to positive interface charge densities of $1 \times 10^{11}\,\mathrm{cm}^{-2}$, $5 \times 10^{11}\,\mathrm{cm}^{-2}$ and $1 \times 10^{12}\,\mathrm{cm}^{-2}$. Figure 3.20 shows the threshold voltage distributions in the fresh transistor and in transistors with three different trapped charge densities. Apart from the average increase in the threshold voltage associated with NBTI degradation in large transistors, a marked change in the slope of the distribution is observed in Fig. 3.20, indicating an increase in the spread with the increase in trap density. Figure 3.20 also clearly indicates that the threshold

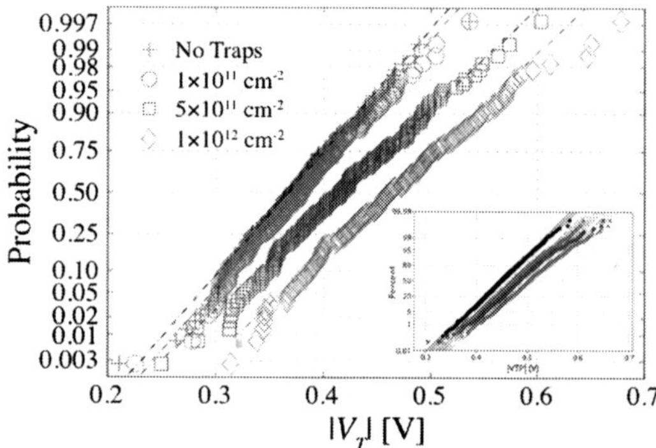

Fig. 3.20 A normal probability plot of the threshold voltages, V_T of 200 microscopically different 45-nm p-channel devices. Plots for devices with no traps, and with trap sheet densities of $1 \times 10^{11}\,\mathrm{cm}^{-2}$, $5 \times 10^{11}\,\mathrm{cm}^{-2}$ and $1 \times 10^{12}\,\mathrm{cm}^{-2}$ are shown. The inset shows experimental results for a similar device [26]

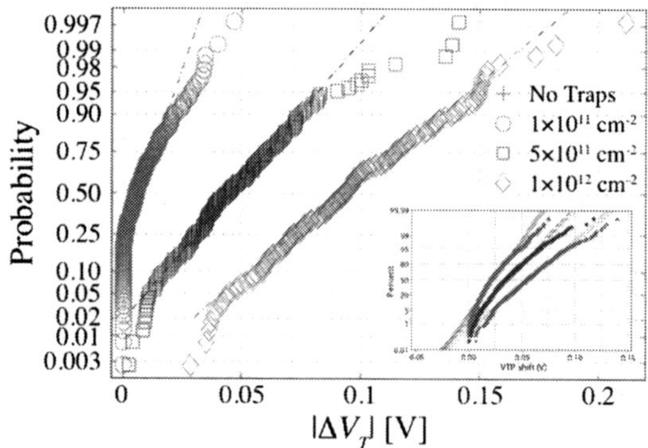

Fig. 3.21 A normal probability plot of the absolute change in threshold voltages, $|\Delta V_T|$ of 200 microscopically different 45-nm p-channel devices with trapped charge at the Si/SiO$_2$ interface. Plots for devices with trap sheet densities of $1 \times 10^{11}\,\mathrm{cm}^{-2}$, $5 \times 10^{11}\,\mathrm{cm}^{-2}$ and $1 \times 10^{12}\,\mathrm{cm}^{-2}$ are shown. The inset shows experimental results for a similar device [26]

voltage distributions are not strictly normal (which would present as a straight line on the plot), and the departure from normality is larger in the tails.

For each of the 200 devices simulated, the change in V_T as a result of the NBTI degradation is shown on a normal probability plot in Fig. 3.21 for the same three levels of degradation. For low trap sheet density (1×10^{11} cm^{-2}), there are many devices that show no increase in threshold voltage at all. In these cases, the traps (if there are any) are not occurring at locations within the channel where there is significant current flow, as dictated by the distribution of discrete impurities. The increase in the density of the trapped charge increases the spread in threshold voltage and hence the slope of the distributions. Figure 3.19 illustrates the most extreme case from Fig. 3.21 for trap density 5×10^{11} cm^{-2} (i.e., the square furthest to the right).

3.8 Variability in Future Technology Generations

In order to foresee the expected magnitude of statistical variability in the future, we have studied the individual impact of RDD, LER and PSG on MOSFETs with gate lengths 35 nm, 25 nm, 18 nm, 13 nm and 9 nm physical gate length. We also compare the results with the statistical variability introduced in the same devices when RDD, LER and PSG are introduced in the same devices simultaneously. The scaling of the simulated devices is based on a 35-nm MOSFET published by Toshiba [27] against which our simulations were carefully calibrated. The scaling closely follows the prescriptions of the ITRS [28] in terms of EOT, junction depth, doping and supply voltage. The intention was also to preserve the main features of the reference 35-nm MOSFET and, in particular, to keep the channel doping concentration at the interface as low as possible. Figure 3.22 shows the structure of the scaled devices. More details about the scaling approach and the characteristics of the scaled devices may be found in [20].

Figure 3.23 compares the channel length dependence *of* σV_T introduced by random dopants, LER and poly-Si grain boundaries with Fermi level pinning. The average size of the polysilicon grains was kept at 40 nm for all channel lengths. Two scenarios for the magnitude of LER were considered in the simulations. In the first scenario, the LER values decrease with the reduction of the channel length following the prescriptions of the ITRS 2003 of 1.2, 1.0, 0.75, and 0.5 nm for the

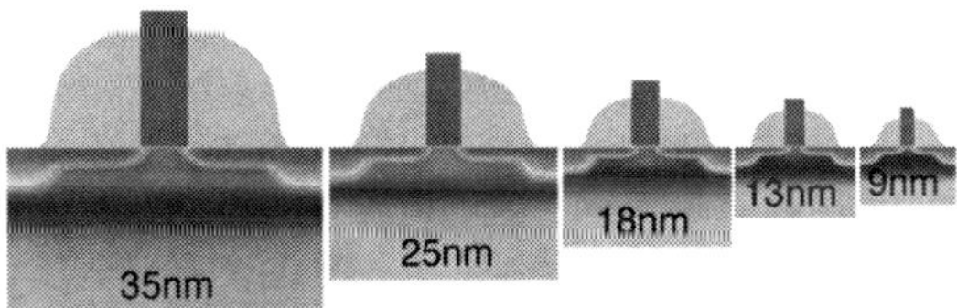

Fig. 3.22 Examples of realistic conventional MOSFETs scaled from a template 35-nm device according to the ITRS requirements for the 90-nm, 65-nm, 45-nm, 32-nm and 22-nm technologies, obtained from process simulation with Taurus Process

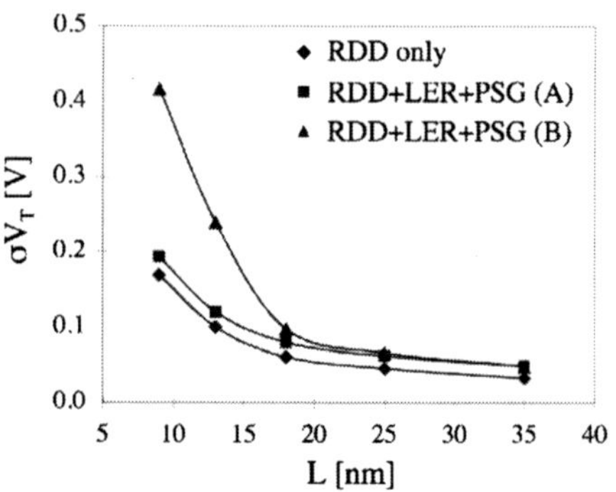

Fig. 3.23 Channel length dependence *of σV_T* introduced by random dopants, line edge roughness and poly-Si granularity: (**a**) LER scales according ITRS; (**b**) LER = 4 nm

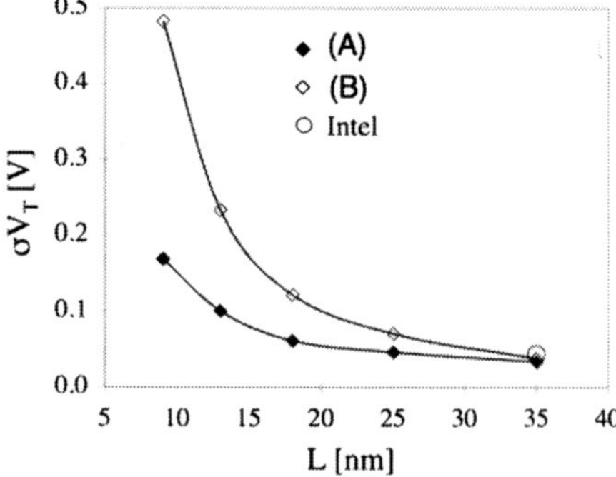

Fig. 3.24 Channel length dependence of σV_T introduced by random dopants, line edge roughness and poly-Si granularity: (**a**) t_{ox} scales according ITRS; (**b**) $t_{ox} = 1$ nm

35-, 25-, 18-, and 13-nm channel length transistors, respectively. In this case the dominant source of variability at all channel lengths is the RDD. The variability introduced by the polysilicon granularity is similar to that introduced by RDD for the 35-nm and 25-nm MOSFETs, but at shorter channel lengths the random dopants take over. The combined effect of the three sources of variability is also shown in the same figure. In the second scenario, LER remains constant and equal to its current value of approximately 4 nm ($\Delta = 1.33$ nm). The results for the 35-nm and the 25-nm MOSFETs are very similar to the results with scaled LER but below 25-nm channel length, LER rapidly becomes the dominant source of variability.

Figure 3.24 is analogous to Fig. 3.23 exploring the scenario of the oxide thickness, which is difficult to scale further. The LER is scaled according to the ITRS requirements listed above. Even with the introduction of high-k gate stack, it is likely to remain stagnated at 1 nm. This will lead to an explosion in the threshold voltage variability for bulk MOSFETs with physical channel length below 25 nm.

The analysis of the numerical simulations shows that the RDD-induced standard deviation of the threshold voltage is proportional to the channel doping concentration to power 0.4 according to the empirical (3.2) [20, 29].

$$\sigma V_T = 3.19 \times 10^{-8} \frac{(t_{ox} + z_0 \varepsilon_{ox}/\varepsilon_{Si})N_A^{0.4}}{\sqrt{L_{eff} W_{eff}}}, \tag{3.2}$$

where t_{ox} is the gate dielectric thickness, z_0 is the inversion layer charge centroid, ε_{ox} and ε_{si} are the dielectric constants of the gate dielectric and the silicon, N_A is the channel doping concentration, and L_{eff} and W_{eff} are the effective channel length and width, respectively.

The scaling of the conventional bulk MOSFETs required continuous increase of the channel doping in order to control the short channel effects and the related leakage. This in turn keeps the RDD-related variability high. Thin body SOI transistors, due to better electrostatic integrity and short channel control, tolerate very low channel doping and, therefore, are resilient to the main source of statistical variability in bulk MOSFETs the RDD. At the same time, very good electrostatic integrity and corresponding reduction of the threshold voltage sensitivity on channel length and drain voltage also reduce their susceptibility to LER-induced variability.

Table 3.3 illustrates the individual and combined impact of RDD, LER and PSG on σV_T in 32-nm UTB SOI and 22-nm double gate MOSFETs. The values of σV_T in the 32-nm UTB SOI and 22-nm double gate (DG) MOSFETs are 3–4 times smaller compared to the equivalent values in bulk MOSFETs with similar size. The results for a 22-nm double gate MOSFET are also representative for FinFET type device architectures.

The reduction of the RDD, LER and PSG variability in the UTB SOI and DG MOSFETs focuses the attention on NBTI-, PBTI- and HCI- related variability. Also, the introduction of high-k gate dielectric and the corresponding relatively high density of fixed and trapped charge (FTC) introduces unwanted variability which can neutralize the benefits from low channel doping and reduced short channel effects. Figure 3.25 illustrates the impact of FTC with different interface charge density on the potential distribution in 32-nm UTB SOI MOSFETS

Table 3.3 Individual and combined impact of RDD, LER and PSG on σV_T in 32 nm UTB SOI and 22 nm double gate MOSFETs

	45 nm σV_T (mV)		32 nm σV_T (mV)		22 nm σV_T (mV)	
	V_{ds} (50 mV)	V_{ds} (1.1 V)	V_{ds} (50 mV)	V_{ds} (1.0 V)	V_{ds} (50 mV)	V_{ds} (1.0 V)
RDD	50	52	5.3	6.1	6.4	8.1
LER	20	33	3.3	8.6	5.8	13
PPG	30	26	N/A	N/A	N/A	N/A
Combined	62	67	6.2	11	8.6	15
Measured	62	69	N/A	N/A	N/A	N/A

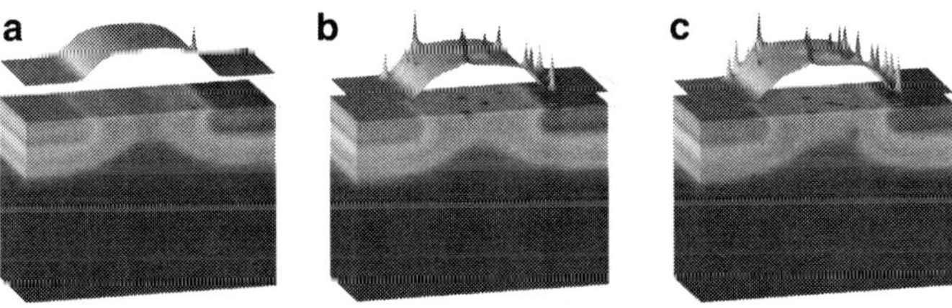

Fig. 3.25 Typical potential profiles corresponding to trap charge with sheet density at (**a**) 1e11 cm^{-2}, (**b**) 5e11 cm^{-2} and (**c**) 1e12 cm^{-2}

described in detail elsewhere [30]. The impact of different interface trapped charge density on σV_T in 32-nm UTB SOI and 22-nm double gate MOSFETs is summarized in Table 3.4.

The 2008 "update" of the ITRS has introduced much more profound changes in the definition of technology generations and the corresponding physical gate length of transistors compared to the 2007 full "edition". Some of these changes have been motivated to great extent by concerns about statistical variability. The dramatic disparity between the number in nm identifying the technology generation (which itself is now divorced from the definition of the half pitch and has become a purely commercial pointer) and the physical gate length, present in the 2007 ITRS edition, practically disappears at the 22-nm technology generation in the 2008 ITRS update. Figure 3.26 illustrates this trend. Bearing in mind that the 2008 ITRS update extends the life of bulk MOSFETS to 2016 and looking into the variability explosion below 25-nm physical channel length in Figs. 3.23 and 3.24, it becomes immediately obvious what is one of the main motivation for such drastic return to realistic physical channel lengths.

In the past, the research in new gate stack materials and new device architectures has been mainly motivated by the drive to improve the device performance. Not any

Table 3.4 Summary of simulation results with different interface charge denisty

	32 nm σV_T (mV)		22 nm σV_T (mV)	
	V_{ds} (50 mV)	V_{ds} (1.0 V)	V_{ds} (50 mV)	V_{ds} (1.0 V)
Trap ($1e11$ cm^{-2})	11	11	5.1	4.8
Trap ($5e11$ cm^{-2})	18	17	13	12
Trap ($1e12$ cm^{-2})	26	23	18	17

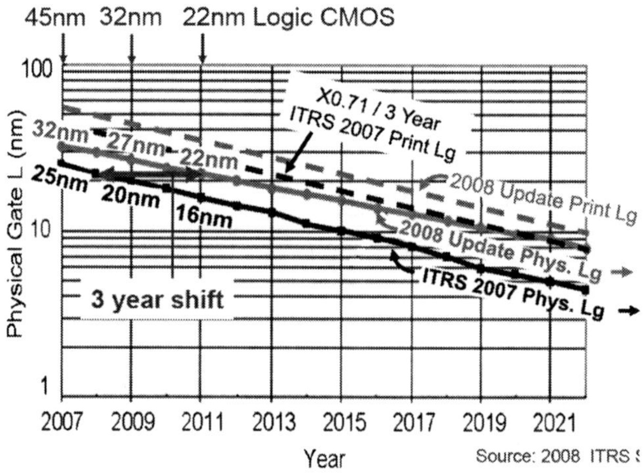

Fig. 3.26 Scaling of the printed and the physical gate length according to the 2008 ITRS update. *Source:* 2008 ITRS Summer Public Conference

more. As illustrated in Fig. 3.27, the main driving force behind the introduction metal gate technology, fully depleted SOI and FinFET devices is the promise for reduction of the statistical variability. However this analysis is based on the assumption that the main source of statistical variability is the RDD. The results of the numerical simulations presented earlier in this section clearly illustrate that due to uncertainties in the future technology solutions and capabilities, Fig. 3.27 represents at this moment of time only wishful thinking. As Fig. 3.23 illustrates, if LER remains at its current level of 4–5 nm at 20-nm physical channel length, it will become the main source of statistical variability in bulk MOSFETs. LER is mainly determined by the physics and chemistry of the 193-nm photolithography process which has been struggling to deliver the required reduction in the feature size for few technology generations. The lack of clear successor of the 193-nm lithography questions the lithography capability to deliver the much needed reduction of the LER.

The introduction of high-k/metal gate stack by Intel at the 45-nm technology generation indeed reduced the statistical variability [31]. This is due to two important factors. Firstly, according to (3.2), the threshold voltage variability is proportional to the oxide thickness and the high-k dielectric has resulted in reduction of the EOT. Secondly, the metal gate removes the polysilicon depletion layer and improves the screening of the RDD-induced potential fluctuations in the channel by the high electron concentration in the gate. The question is to what extent the technology will be able to reduce the EOT below the current 1 nm mark. Figure 3.24, which is benchmarked against the measured variability in the Intel 45-nm technology, indicates that if the EOT reduction, planned by the ITRS, is achieved, the useful life of the bulk MOSFET from statistical variability point of view may be extended below 20-nm channel length. However, if the EOT cannot be scaled further, there will be great difficulties to achieve reasonable statistical variability in bulk MOSFETs at the 22-nm technology generation. For example, an important limitation of the EOT scaling is the presence of almost invariable distance between the charge centroid of the channel and the interface (z_0 in (3.1)) due to the quantum confinement effects. Also, the introduction of high-k/metal gate has its own problems from variability point of view. Although it reduces the EOT, it introduces new variability sources compared to the old

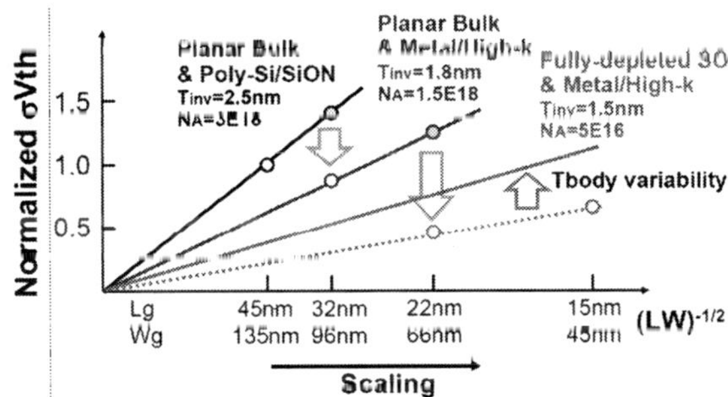

Fig. 3.27 Statistical variability reduction scenario. *Source*: 2007 ITRS Winter Public Conference

fashioned SiO_2 gate stack technology. First of all, the high-k dielectric has typically a higher density of fixed/trapped charges. As illustrated in Fig. 3.21 and Table 3.4, this can significantly increase the variability in fully depleted SOI MOSFETs. More than this, the p-channel high-k transistors are more susceptible to NBTI which can cause increase in the statistical variability ageing. This picture is exacerbated by the creeping PBTI [32] in the n-channel high-k transistors which was insignificant in their SiO_2 gate stack counterparts. The granularity of the high-k dielectric and the metal gate has also become a concern from the statistical variability point of view [11, 33].

3.9 Compact Model Strategies for Statistical Variability

It is very important to be able to capture the simulated or measured statistical variability in statistical compact models since this is the only way to communicate this information to designers [34]. Previous research on statistical compact model identification was focused mainly on variability associated with traditional process variations resulting from poor control of critical dimensions, layer thicknesses and doping clearly related to specific compact model parameters [35, 36].

Unfortunately, the current industrial strength compact models do not have natural parameters designed to incorporate seamlessly the truly statistical variability associated with RDD, LER, PGG and other relevant variability sources. Despite some attempts to identify and extract statistical compact model parameters suitable for capturing statistical variability introduced by discreteness of charge and matter, this remains an area of active research [37]. Figure 3.28 shows the spread in I_D–V_G characteristics obtained from 'atomistic' simulator due to the combined effect of RDD, LER and PGG.

We use the standard BSIM4 compact model [38] to capture the information for statistical variability obtained from full 3D physical variability simulation.

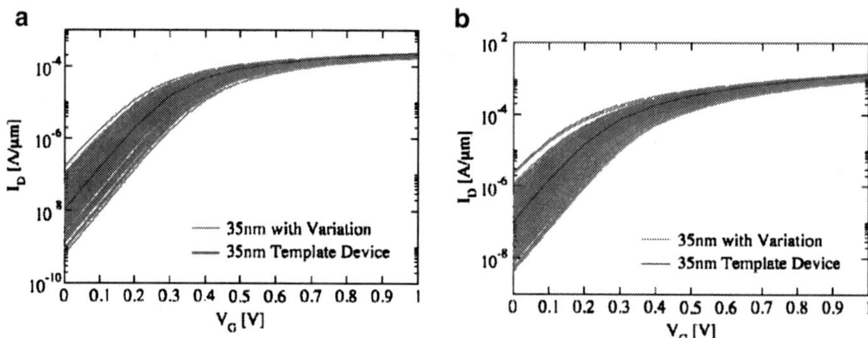

Fig. 3.28 Variability in the current voltage characteristics of a statistical sample of 200 microscopically different 25-nm square (W = L) n-channel MOSFETs at (**a**) V_D = 50 mV and (**b**) V_D = 1 V

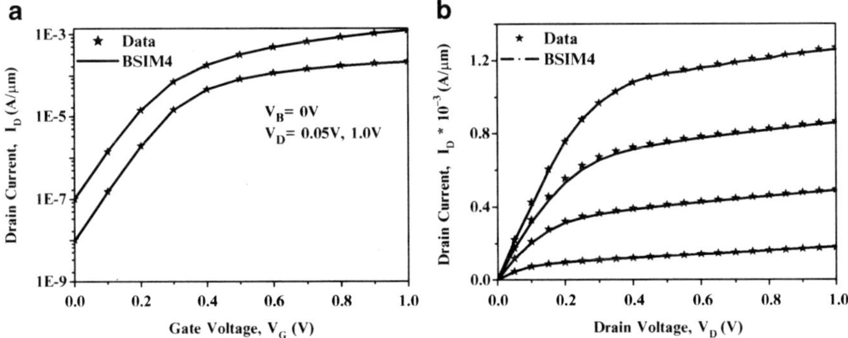

Fig. 3.29 Comparison of (**a**) I_D–V_G characteristics, (**b**) I_D–V_D characteristics for 'uniform' devices obtained from physical simulations and BSIM

The statistical extraction of compact model parameters is done in two stages [39]. In the first stage, one complete set of BSIM4 parameters is extracted from the I–V characteristics of "uniform" (continuously doped, no RDD, LER and PGG) set of devices with different channel lengths and widths, and process flow identical to that of one of the 35-nm testbed transistor [20]. Target current voltage characteristics are simulated over the complete device operating range, and parameter extraction strategy combining group extraction and local optimization is employed. Figure 3.29 compares the corresponding BSIM4 generated I–V characteristics of the 35-nm transistor with the original device characteristics obtained from our 'atomistic' simulator. The RMS error for this extraction is 2.8%.

At the second stage, we re-extracted a small carefully chosen subset of the BSIM4 model parameters from the physically simulated characteristics of each microscopically different device in the statistical ensemble keeping the bulk of the BSIM parameters unchanged. The transfer (I_D–V_G) characteristics at low and high drain bias are used as extraction target at this stage. The seven re-extracted model parameters are L_{pe0}, R_{dswmin}, N_{factor}, V_{off}, A_1, A_2 and D_{sub}.

Figure 3.30 compares the BSIM4 results for the worst, the best and the typical devices with the physically simulated device characteristics. The good match between the BSIM results and the physically device characteristics validates the choice of seven parameters, which guaranty high accuracy of the compact model over the whole statistical range of physically simulated device characteristics. Figure 3.31 illustrates the distribution in the error of the 200 statistically different BSIM4 cards depending on the density of the data points in the I_D–V_G characteristics used as targets in the statistical compact model extraction. Relatively few simulation data points are needed to extract high-accuracy statistical compact model parameters.

The scatter plots in Fig. 3.32 show that the chosen seven BSIM4 parameters are not all statistically independent and, therefore, cannot be generated independently. Also, as shown in Fig. 3.33, the distribution of most of the seven parameters is not necessarily a normal distribution, which means that it would be difficult to generate

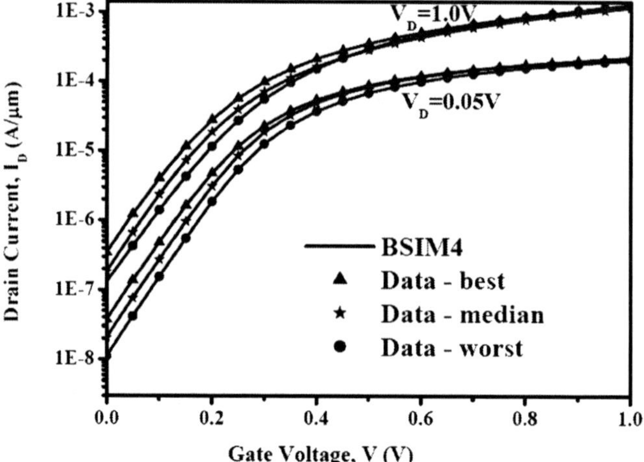

Fig. 3.30 Comparison of physical and simulation results obtained after second stage statistical extraction for the devices with best, worst and median extraction error

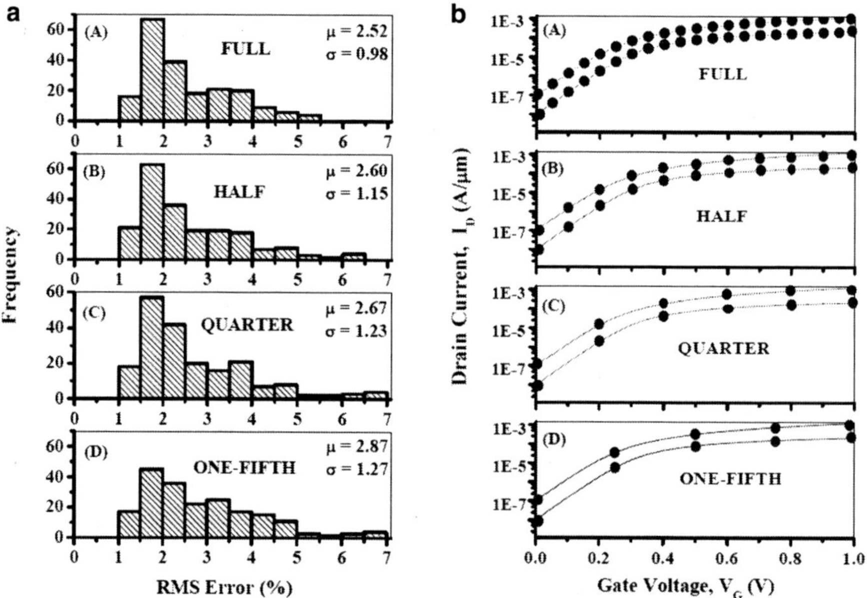

Fig. 3.31 Distribution of the statistical error as a function (**a**) for a different number of data points on the target I_D–V_G characteristics used in the statistical compact model extraction

statistically accurate parameter sets by employing ordinary multivariate statistical methods, such as principal component analysis. In order to reproduce precisely each individual simulated device, a statistical compact model library is built and

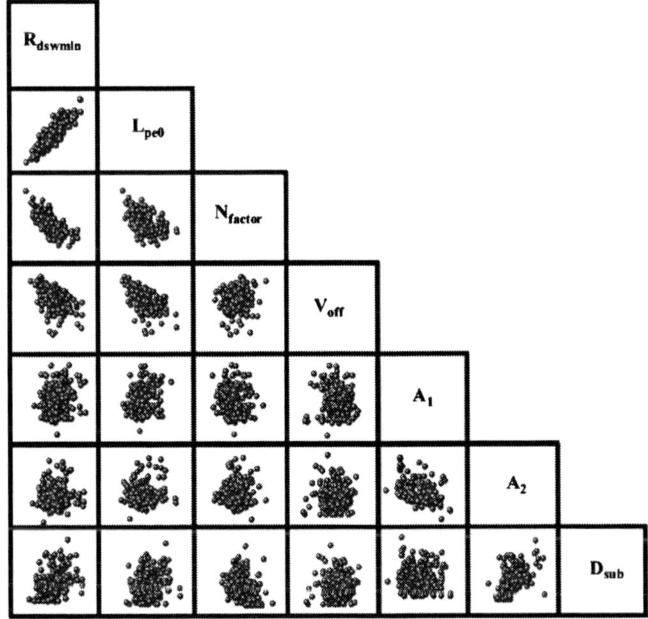

Fig. 3.32 *Scatter plots* between each pair of the statistical compact model parameters

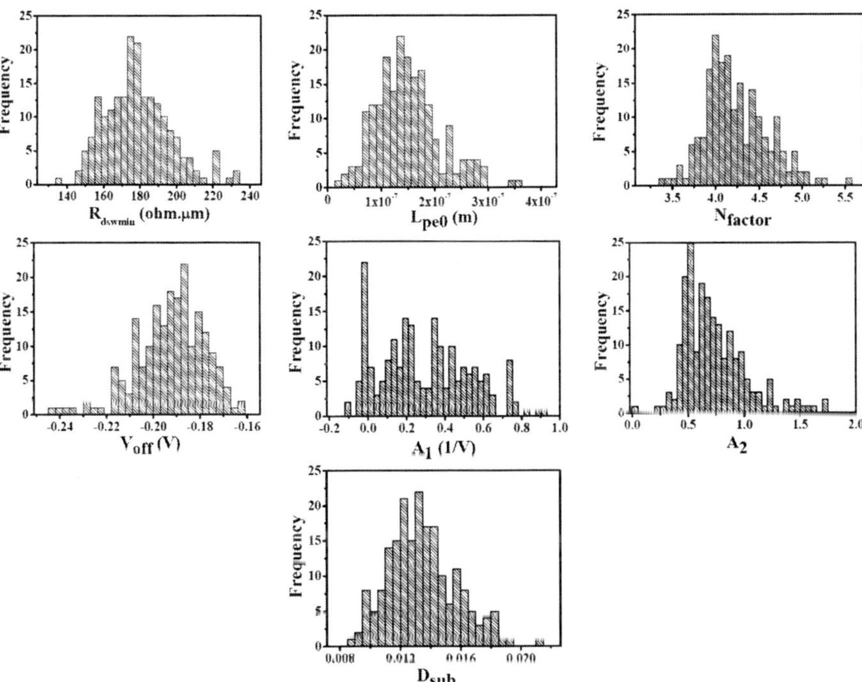

Fig. 3.33 Distribution of the seven statistical compact model parameters

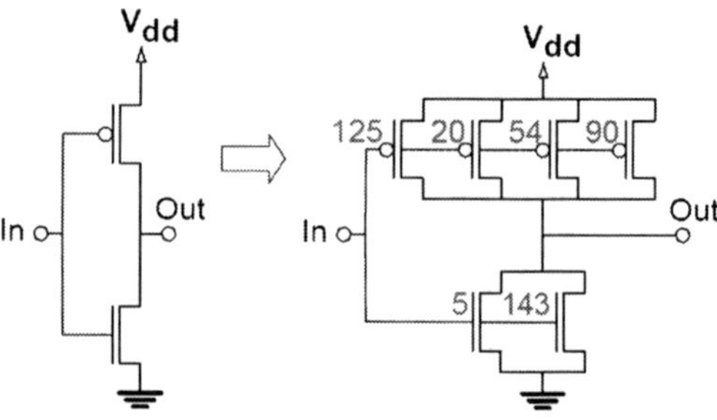

Fig. 3.34 Schematic of construction of multi-width/length ratio devices, using inverter with W/L of 4 for PMOS, W/L of 2 for NMOS as an example

statistical instances of devices in statistical circuit simulation can be randomly selected from the library. Statistical weight can also be assigned to each of the statistical compact models for the purpose of statistical enhancement of the statistical circuit simulation.

The above statistical compact model extraction and 3D physical device simulation are based on minimum size square device ($W = L$), which has the maximum magnitude of statistical variation. However, in real circuits, most of the devices have multi-width/length ratio value. In order to reproduce the right statistical behavior of such devices in statistical circuit simulation, wider devices are constructed by connecting in parallel minimum-size devices randomly selected from the statistic compact model library. This is illustrated in Fig. 3.34, which shows schematically the construction of multi-width/length ratio devices, using inverter with W/L of 4 for PMOS and W/L of 2 for NMOS as an example. The numbers in the right side circuit illustrate the randomly selected compact model cards from the available sample of 200 compact models with seven different statistical parameters corresponding to the 200 physically simulated microscopically different transistor characteristics from Fig. 3.28. The drawback of this approach is the increased number of transistors in the Spice simulation. This number can be reduced by re-extraction of the seven-parameter set from the spice-simulated I–V characteristics of the composition transistors.

3.10 Basics of Statistical Circuit Simulations in the Presence of Statistical Variability

The statistical circuit simulation methodology described in the previous section, which can transfer all the fluctuation information obtained from 3D statistical device simulations into circuit simulation, is employed to investigate the impact of RDF on 6T and 8T SRAM stability for the next three generations of bulk CMOS technology. In the following discussions, we use 25-nm, 18-nm and 13-nm channel length transistors described in detail in [14].

Currently, 6T SRAM is the dominant SRAM cell architecture in SoC and microprocessors. However, the disturbance of bit lines on the data retention

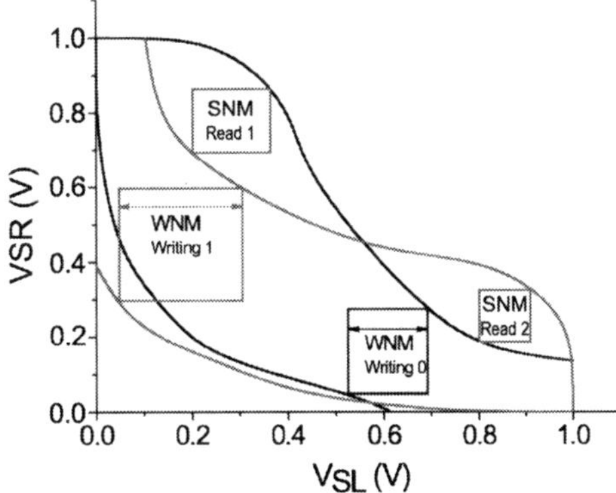

Fig. 3.35 Static voltage transfer characteristics and definition of SNM and WNM

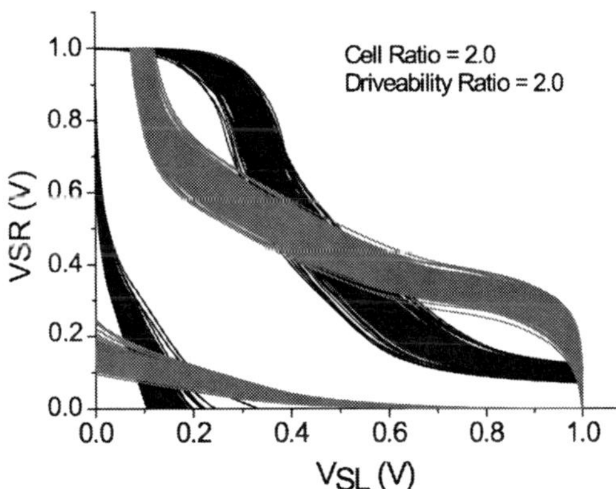

Fig. 3.36 Statistical behavior of the SNM and WNM of SRAM made of 25-nm bulk MOSFETs subject to RDD [40]

element during read access makes the 6T cell structure vulnerable to statistical variability, which in turn will have a huge impact on 6T SRAM's scalability. The functionality of SRAM is determined by both static noise margin (SNM) defined as the minimum DC voltage necessary to flip the state of the cell and the write noise margin (WNM) defined as the DC noise voltage needed to fail to flip a cell during a write period. The meaning of SNM and WNM is defined in Fig. 3.35. Figure 3.36 illustrates the statistical nature of SNM and WNM in the presence of statistical transistor variability. The magnitude of WNM in SRAM is mainly determined by the load and access transistors illustrated in the 6T SRAM schematics inset in Fig. 3.37. Since they are the smallest transistors in an SRAM cell, the WNM variation will be larger than the SNM variation. However, the mean value of WNM is much larger than its SNM counterpart. Previous studies [41] suggested that under normal circumstances, the limiting factor for the operation of bulk 6T SRAM cells is SNM.

The statistical circuit simulation results for the distribution of SNM for 6T SRAM with the cell ratio of 2 are shown in Fig.3.37, while the schematic of a 6T

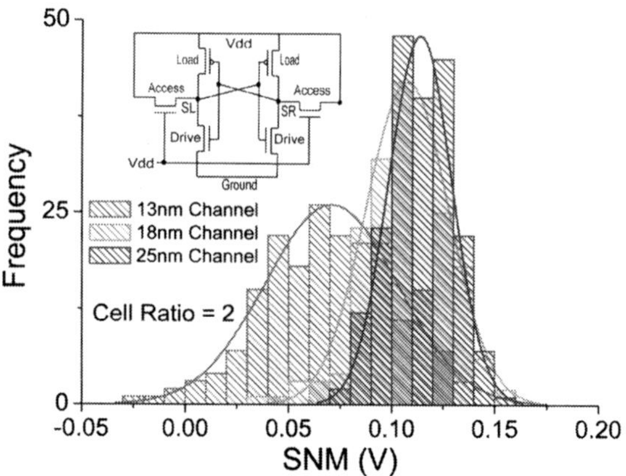

Fig. 3.37 Distribution of 6 T SRAM SNM over an ensemble of 200 SRAM cells for 25-, 18- and 13-nm generations

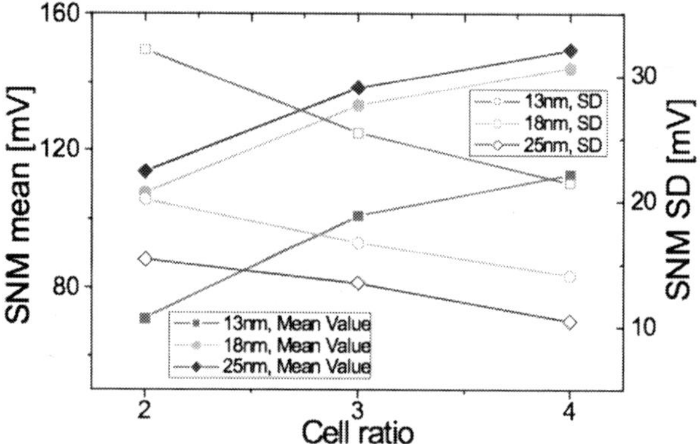

Fig. 3.38 Mean value and SD of 6 T SRAM SNM against SRAM cell ratio

SRAM cell with the bias configuration at the initiation of read operation is illustrated in the inset. The cell ratio is defined as the ratio of the driver transistor to access resistor channel widths. For the 13-nm generation, around 2% of SRAM cells with cell ratio of 2 are not readable even under ideal conditions since their SNM values are negative. The standard deviation σ of SNM, normalized by the average value μ of the SNM, increases from 13% for the 25-nm generation, to 19% for the 18-nm generation, and reaches 45% for the 13-nm generation. As a guideline, μ-6σ is required to exceed approximately 4% of the supply voltage to achieve 90% yield for 1Mbit SRAM's [42]. Although the 25-nm generation is three times better with respect to SNM fluctuation performance compared to the 13-nm generation at cell ratio of 2, it still cannot meet the "μ-6σ" test on yield control.

Increasing the cell ratio is the most straightforward way to improve SNM performance of an SRAM cell [39]. The mean value (μ) and the standard deviation (σ) of SNM at different SRAM cell ratios are shown in Fig. 3.38, which clearly illustrates the benefits of a larger cell ratio. For the 25-nm transistors, the increase of the cell ratio from 2 to 4 results in nearly two times NSD of SNM improvement, and the "μ-6σ" criterion is met at a cell ratio of 3. Although the 18-nm and 13-nm generations follow a similar trend, a cell ratio of 4 is required for the 18-nm technology in order to achieve a reasonable yield. For the 13-nm generation, a cell ratio of 4 can only achieve the level of performance of the 18-nm generation at a cell ratio of 2. At the same time, the increase in the cell ratio tends to degrade the WNM [41]. Therefore, the approach of increasing the cell ratio becomes less attractive for extremely scaled devices even from the prospect of the yield.

In order to improve the SRAM cell immunity to statistical variability, new cell structures with read-disturbance free operation will be needed. In contrast to the 6T cell, the access transistors are switched off during a read operation in the 8T cell illustrated in Fig. 3.39, with connections corresponding to the simulation conditions, and the data are passively accessed through transistors M1 and M2, which guarantee that the bit line will not have a direct impact on the stored data.

The results for the distribution of SNM for 8T SRAM with the cell ratio of 2 are shown in Fig. 3.40. The mean μ of SNM is 260 mV, 255 mV and 220 mV, respectively, for 25-, 18- and 13-nm transistors, and the corresponding SNM standard deviations are 0.054, 0.077 and 0.10, respectively. Depending on the

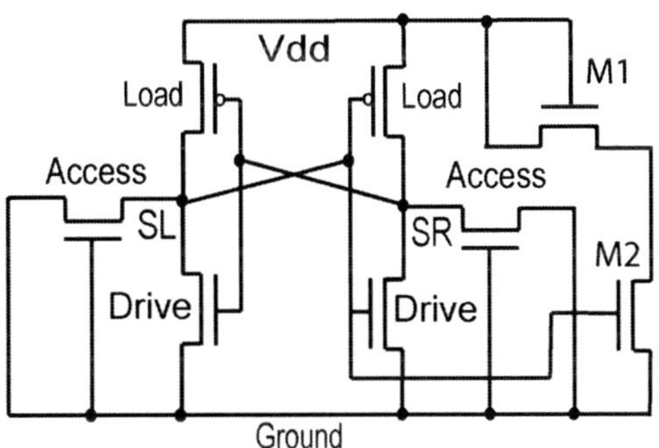

Fig. 3.39 Bias configuration for SNM of 8 T SRAM cell

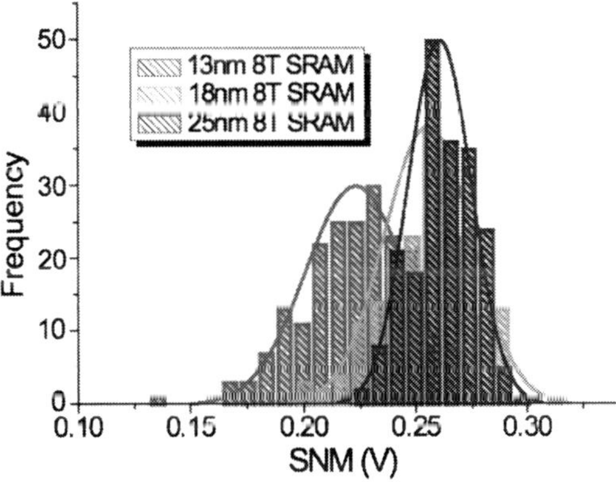

Fig. 3.40 Distribution of 8 T SRAM SNM over an ensemble of 200 SRAM cells for 25-, 18- and 13-nm generations

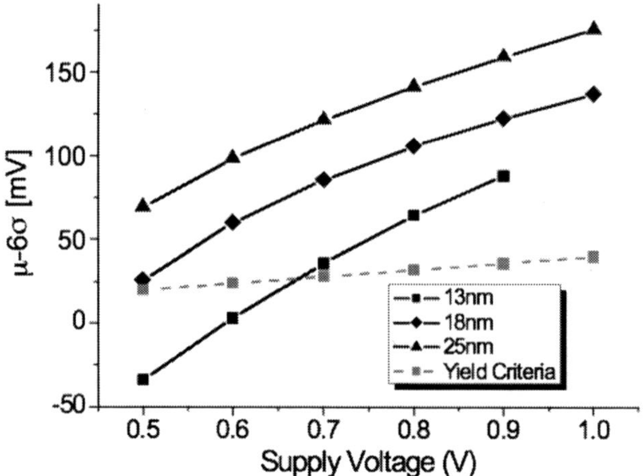

Fig. 3.41 Six sigma yield performance of 8 T SRAM cell against

technology generation, at least 200% variation improvement can be achieved if we shift from a 6T cell to an 8T cell. From an SNM point of view, all three generations of 8T cells can pass the "μ-6σ" yield test by a considerable margin.

One of the fundamental limitations on future integrated systems is power dissipation [2]. However, supply voltage scaling has been historically, and will continue to be, much less aggressive than device scaling, with yield being one of the major reasons behind this. Since 8T SRAM can provide a much improved SNM figure, this in turn could open a window for bolder power supply scaling. However, this will carry the penalty of approximately 20% increase in the cell area. Figure 3.41 shows the yield criterion for 8T SRAM cells with a cell ratio of 2 against supply voltage. For the 25-nm and 18-nm generations, as far as SNM is concerned, reasonable yield can be expected even when the supply voltage drops to 0.5 V. For the 13-nm generation, although the normalized standard deviation of the on-current can be 20% of the mean due to RDF, the minimum workable supply voltage can still remain at 0.7 V. Since the minimum supply voltage of a ULSI system is usually limited by its SRAM component [43], employing an 8T cell structure will provide greater flexibility with respect to power budget issues in system design.

There is no difference in write operation between the 6T and 8T SRAM cells, which indicates that for 8T SRAM – due to the much improved SNM performance – the limiting factor on stability may shift to WNM.

3.11 Conclusions

The statistical variability introduced by discreteness of charge and matter has become one of the major concerns for the semiconductor industry. More and more of the strategic technology decisions that the industry will be making in the future will be motivated by the desire to reduce statistical variability. The useful life of bulk MOSFETs, from the statistical variability point of view, can be extended below the 20-nm technology mark only if the LER and EOT could be successfully scaled to the required values. The introduction of fully depleted SOI MOSFETs and perhaps FinFETs will be mainly motivated by the necessity to reduce the statistical variability. This, however, might be jeopardised by other sources of variability associated with the introduction of high-k/metal gate stack and increased statistical reliability problems. SRAM, which uses minimum channel width transistors, is the most sensitive part of the integrated systems in respect of statistical variability and needs special care and creative design solution in order to take full advantage from scaling in present and future technology generations.

References

1. Moore GE (1975) Progress in digital electronics, technical digest of the Int'l Electron Devices Meeting, IEEE Press, 13
2. Bernstein K, Frank DJ, Gattiker AE, Haensch W, Ji BL, Nassif SR, Nowak EJ, Pearson DJ, Rohrer NJ (2006) IBM J Res Dev 50:433
3. Brown AR, Roy G, Asenov A (2007) Poly-Si gate related variability in decananometre MOSFETs with conventional architecture. IEEE Trans Electron Devices 54:3056
4. Cheng B-J, Roy S, Asenov A (2004) The impact of random dopant effects on SRAM cells. Proc. 30th European Solid-State Circuits Conference (ESSCIRC), Leuven, 219
5. Agarwal A, Chopra K, Zolotov V, Blaauw D (2005) Circuit optimization using statistical static timing analysis. Proc. 42nd Design Automation Conference, Anaheim, 321
6. Kahng AB, Pati YC (1999) Subwavelength optical lithography: challenges and impact on physical design. Proceedings of the 1999 international symposium on Physical design, 112
7. Wang X, Cheng B, Roy S, Asenov A (2008) Simulation of strain induced variability in nMOSFETs. Proceedings of the 9th international conference on ultimate integration on silicon, 15–16 March, Udine, Italy, p 89
8. See for an example: http://www.synopsys.com/Tools/Manufacturing/MaskSynthesis/ WorkBenchPlus/Pages/default.aspx
9. Asenov A (1998) Random dopant induced threshold voltage lowering and fluctuations in sub 0.1 micron MOSFETs: a 3D 'atomistic' simulation study. IEEE Trans Electron Devices 45:2505
10. Asenov A, Kaya S, Brown AR (2003) Intrinsic parameter fluctuations in decananometre MOSFETs introduced by Gate Line Edge Roughness. IEEE Trans Electron Devices 50:1254
11. Watling JR, Brown AR, Ferrari G, Babiker JR, Bersuker G, Zeitzoff P, Zeitzoff P, Asenov A (2008) Impact of high-k on transport and variability in nano-CMOS devices. J Comput Theor Nanosci 5(6):1072
12. Asenov A, Kaya S, Davies JH (2002) Intrinsic threshold voltage fluctuations in decanano MOSFETs due to local oxide thickness variations. IEEE Trans Electron Devices 49:112
13. Brown AR, Watling JR, Asenov A (2002) A 3-D atomistic study of archetypal double gate MOSFET structures. J Comput Electron 1:165
14. Rauch SE III (2007) Review and reexamination of reliability effects related to NBTI-induced statistical variations. IEEE Trans Device Mater Rel 7:524
15. Reddy V et al (2002) Impact of negative bias temperature instability on digital circuit reliability. Proc IEEE IRPS, 248 254
16. Compagnoni CM, Spinelli AS, Beltrami S, Bonanomi M, Visconti A (2008) IEEE Electron Dev Lett 29:941

17. Fischer T, Amirante E, Hofmann K, Ostermayr M, Huber P, Schmitt-Landsiedel D (2008) Proc ESSDERC'08, 51
18. Markov S, Brown AR, Cheng B, Roy G, Roy S, Asenov A (2007) Three-dimensional statistical simulation of gate leakage fluctuations due to combined interface roughness and random dopants. Jpn J Appl Phys 46(4B):2112
19. Rodríguez R, Stathis JH, Linder BP, Kowalczyk S, Chuang CT, Joshi RV, Northrop G, Bernstein K, Bhavnagarwala AJ, Lombardo S (2002) The impact of gate-oxide breakdown on SRAM stability. IEEE Electron Device Lett 23:559
20. Roy G, Brown AR, Adamu-Lema F, Roy S, Asenov A (2006) Simulation study of individual and combined sources of intrinsic parameter fluctuations in conventional nano-MOSFETs. IEEE Trans Electron Devices 52:3063–3070
21. Taiault J, Foucher J, Tortai JH, Jubert O, Landis S, Pauliac S (2005) Line edge roughness characterization with three-dimensional atomic force microscope: transfer during gate patterning process. J Vac Sci Technol B 23:3070
22. Uma S, McConnell AD, Asheghi M, Kurabayashi K, Goodson KE (2001) Temperature-dependent thermal conductivity of undoped polycrystalline silicon layers. Int J Thermophys 22:605
23. Cathignol A, Cheng B, Chanemougame D, Brown AR, Rochereau K, Ghibaudo G, Asenov A (2008) Quantitative evaluation of statistical variability sources in a 45 nm technological node LP N-MOSFET. IEEE Electron Device Lett 29:609
24. Asenov A, Cathignol A, Cheng B, McKenna KP, Brown AR, Shluger AL, Chanemougame D, Rochereau K, Ghibaudo G (2008) Origin of the asymmetry in the magnitude of the statistical variability of n- and p-channel poly-Si gate bulk MOSFETs. IEEE Electron Device Lett 29:913–915
25. Bukhori MF, Roy S, Asenova A (2008) Statistical aspects of reliability in bulk MOSFETs with multiple defect states and random discrete dopants. Microelectron Reliab 48(8–9):1549–1552
26. Huard V, Parthasarathy C, Guerin C, Valentin T, Pion E, Mammasse M, Planes N, Camus L (2008) NBTI degradation: from transistors to SRAM arrays, IEEE CFP08RPS-CDR 46th annual international reliability physics symposium, p 289
27. Inaba S, Okano K, Matsuda S, Fujiwara M, Hokazono A, Adachi K, Ohuchi K, Suto H, Fukui H, Shimizu T, Mori S, Oguma H, Murakoshi A, Itani T, Iinuma T, Kudo T, Shibata H, Taniguchi S, Takayanagi M, Azuma A, Oyamatsu H, Suguro K, Katsumata Y, Toyoshima Y, Ishiuchi H (2002) High performance 35 nm gate length CMOS with NO oxynitride gate dielectric and Ni salicide. IEEE Trans Electron Devices 49:2263
28. International Roadmap for Semiconductors (ITRS), Semiconductor Industry Association. Available: http://www.itrs.net
29. Asenov A, Saini S (1999) Supression of random dopant induced threshold voltage fluctuations in sub-0.1 µm MOSFETs with epitaxial and delta doped channels. IEEE Trans Electron Devices 46(8):1718–1724
30. Cheng B, Roy S, Brown AR, Millar C, Asenov A (2009) Evaluation of statistical variability in 32 and 22 nm technology generation LSTP MOSFETs. Solid State Electron 53(7):767–772
31. Kuhn KJ (2007) Reducing variation in advanced logic technologies: approaches to process and design for manufacturability of nanoscale CMOS. IEDM Tech Dig, 471
32. Denais M, Huard V, Parthasarathy C, Ribes G, Perrier F, Revil N, Bravaix A (2004) Interface trap generation and hole trapping under NBTI and PBTI in advanced CMOS technology with a 2-nm gate oxide. IEEE Trans Device Mater Rel 4:715
33. Dadgour H, De V, Banerjee K (2008) Statistical modeling of metal-gate work-function variability in emerging device technologies and implications for circuit design. Proc. 2008 IEEE/ACM International Conference on Computer-Aided Design, p 270
34. Lin C-H, Dunga MV, Lu D, Niknejad AM, Hu C (2008) Statistical compact modeling of variations in nano MOSFETs VLSI-TSA 2008, p 165
35. Power JA, Donellan B, Mathewson A, Lane WA (1994) Relating statistical MOSFET model parameter variabilities to IC manufacturing process fluctuations enabling realistic worst case design. IEEE Trans Semicond Manufact 7:306

36. McAndrew CC (2004) Efficient statistical modeling for circuit simulation. In: Reis R, Jess J (eds) Design of systems on a chip: devices and components, Kluwer Academic, p 97
37. Takeuchi K, Hane M (2008) Statistical compact model parameter extraction by direct fitting to variations. IEEE Trans Electron Devices 55:1487
38. BSIM4 manual. Available: http://www-device.eecs.berkeley.edu/
39. Cheng B, Roy S, Roy G, Adamu-Lema F, Asenov A (2005) Impact of intrinsic parameter fluctuations in decanano MOSFETs on yield and functionality of SRAM cells. Solid State Electron 49:740
40. Cheng B, Roy S, Asenov A (2006) Low power, high density CMOS 6-T SRAM cell design subject to 'atomistic' fluctuations. Proc ULIS 2006, ISBN:88-900874-0-8, pp 33–36, Grenoble
41. Cheng B, Roy S, Asenov A (2007) CMOS 6-T SRAM cell design subject to 'atomistic' fluctuations. Solid State Electron 51:565
42. Stolk PA, Tuinhout HP, Duffy R, Augendre E, Bellefroid LP, Bolt MJB, Croon J et al (2001) CMOS device optimization for mixed-signal technologies. Tech Dig IEDM, 215–218
43. Takeda K, Hagihara Y, Aimoto Y et al (2005) A read-static-noise-margin-free SRAM cell for low-VDD and high-speed CMOS SRAM cells. Tech Dig IEDM, 659–662

Chapter 4
Importance Sampling-Based Estimation: Applications to Memory Design

Rouwaida Kanj and Rajiv Joshi

The odds of winning the lottery are 1 in 3.8 million. Same are the desired odds of a memory fail. (Anonymous)

4.1 Introduction

With the technology scaling, aggressive design rules and millions of transistors on a chip, guarding against rare fail events is becoming an indispensable part of the memory design flow. While traditional statistical sampling methods like typical Monte Carlo can be employed to estimate fail probabilities of a design, these techniques can be too slow at predicting probabilities of the order of one in a few million. Hence, there is a need for faster and more efficient statistic-based methodologies. In this chapter, we introduce the concept of importance sampling and its applications to memory design yield prediction. Similar to Monte Carlo methods, importance sampling is a statistical sampling based methodology. However, it is tailored to estimate extremely low fail probabilities efficiently.

The chapter is organised as follows. Section 4.2 discusses statistical sampling in general and random sample generation. Section 4.3 provides a brief history of Monte Carlo methods, its application to numerical integration and statistical inference. It also describes the problem of using Monte Carlo for rare fail events. Section 4.4 provides a brief overview of variance reduction statistical methods including importance sampling methods. Finally, Sect. 4.5 discusses importance sampling for memory designs.

R. Kanj (✉)
IBM, Austin Research Labs, Austin, TX, USA

A. Singhee and R.A. Rutenbar (eds.), *Extreme Statistics in Nanoscale Memory Design,* Integrated Circuits and Systems, DOI 10.1007/978-1-4419-6606-3_4,

4.2 Statistical Sampling

4.2.1 Probability and Statistics: A History in the Making

When John Graunt wrote his book on the *Natural and Political Observations upon the Bills of Mortality* in 1662, he was not aware that it will be considered by scholars to be the origin of a new and distinct mathematical science – later called Statistics [1]. Little did he know that on the other shores across the English channel, two French mathematicians, Blaise Pascal and Pierre de Fermat [2], were laying the mathematical foundations for this science over a gambling dispute. In fact, probability theory was established to answer the famous 1654 dispute on the chances of double six in 24 dice throws. In the twentieth century, probability theory that rose from the games of chance evolved into the more modern measure theory.

4.2.2 Overview of Sampling Methods

> *To the philosopher, this procedure has no justification whatsoever because representative sampling is not how truth is pursued in philosophy. To the scientist, however, representative sampling is the only justified procedure for choosing individual objects for use as the basis of generalization, and is therefore usually the only acceptable basis for ascertaining truth.* (Andrew A. Marino) [3].

According to the Merriam-Webster dictionary, a census is a complete study of the population. This can be expensive, time consuming and often impractical. Sometimes the whole population is not tangible, and a representative sample is sought. Sampling can be over any dimension in terms of time, space and population core characteristics. Sampling techniques vary based on the sample selection methodology. The most common sampling methods are as follows [4]:

- Simple random sampling (SRS): Each sample point is chosen randomly and entirely by chance. Unbiased random sampling implies that the sample points represent the population. However, it is still vulnerable to sampling error. For example, for ten coin tosses, we cannot guarantee observing five heads and five tails. Systematic and stratified sampling aims at sample points that are more representative of the population.
- Systematic sampling: Also called interval sampling, it starts with an ordered population. The first sample point is random, and the remaining sample points are at fixed intervals; e.g. every tenth name in a mailing list.
- Stratified sampling: The population is divided into different groups or strata. Each group is then randomly sampled. This way the sample points are more representative of the population.

- Cluster sampling: The population is divided into groups or clusters. A number of clusters are selected randomly. All elements in a cluster are assumed to be sample points. This is different from stratified sampling where the target is to sample from all the strata.

Other sampling techniques include quota sampling, convenience sampling, etc. We refer the reader to [4] for further details.

4.2.3   Random Sample Generation

Most statistical sampling methods invoke random sample generators. In this section, we review software-based methods for generating uniform random samples and Gaussian random samples.

4.2.3.1   Pseudo-Random Number Generation

A true random number generator would rely on a physical process like coin toss or a dice throw. Software-generated random numbers are often referred to as *apparently* random numbers or "pseudo"-random numbers because they are generated from deterministic processes or computer programs. In fact for the same seed, the random numbers are nothing but a deterministic sequence with good statistical properties.

Park and Miller [5] state that in order "to write quality software which produces what is really desired," pseudo-random generators must generate "a virtually infinite sequence of statistically independent random numbers, uniformly distributed between 0 and 1. This is a key point: strange and unpredictable is not necessarily random." In the Art of Computer Programming, D. Knuth [6] discusses different assorted standard tests for probing for non-randomness in the output of pseudo-random number generators. Practically, high-quality random numbers must have a very large repeat period. Failing the statistical tests can be due to correlation between consecutive sample points, poor behavior in high-dimensional space and many other factors like lacking the basic assumption of the uniformity of the sample points. Finally, the seed determines the arbitrary starting point in the sequence.

Most routine libraries have standard built-in functions that generate uniform random numbers (e.g. rand() in the standard C library). It is important to focus on identifying simple and good random number generators that are portable to all systems [5]. Linear congruential generators (LCG) are among the oldest random number generators. They are based on linear congruential multipliers. An example pseudo-random sequence of a LCG is defined by (4.1).

$$x_{n+1} = (ax_n + c) \bmod m \tag{4.1}$$

$$c = 0 \quad a = 16,807 \quad \text{and} \quad m = 2^{31} - 1 \tag{4.2}$$

x_j is the jth sample point, a is the multiplier and m is the prime modulus. The period of the random number generator is large: $2^{31} - 2 \cong 2 \times 10^9$. A straight forward implementation in a high-level language would be as follows.

```
Pseudo-code: LCG

a=16807
m=2147483647
seed=(a×seed) mod m
rand= seed/m
return rand
```

One would normally need to provide an initial seed value. The problem with this implementation is that it requires a 64-bit system. Additional manipulations are needed to implement the algorithm on a 32-bit system [5]. In general, LCG are suitable for systems with low memory requirements, but may raise some issues when dealing with high-accuracy problems. A more sophisticated random number generator algorithm is the Mercenne Twister algorithm by Matsumoto and Nishimura [7]. It fixes many of the problems encountered in the previous algorithm. It is characterised by a very large period and offers around 10^{6001} unique combinations. It has been shown to be equi-distributed upto 623 dimensions. Variants have been adopted by state-of-the-art processors.

Non-uniform random numbers can be derived from uniform random numbers and the corresponding cumulative distribution functions. In the following section, we will discuss Gaussian random number generation.

4.2.3.2 Generating Normally Distributed Random Numbers

The Gaussian distribution with its bell-shaped probability density function is a simple distribution often used to model complex phenomena. The central limit theorem states that the distribution of the sum of independent identically distributed random variables with finite mean and standard deviations tends to a Gaussian random variable [8] as the number of summands increases. Equation (4.3) lists the probability density function (pdf) of a Gaussian random variable with mean μ and standard deviation σ. We say that x is normally distributed: $x \sim N_g(\mu,\sigma^2)$.

$$P(x) = \frac{1}{\sigma\sqrt{2\Pi}} \exp\left(-\frac{(x-\mu)^2}{2\sigma^2}\right) \tag{4.3}$$

According to the fundamental laws of probabilities, if y is a uniform random variable $\in [0, 1]$, then

$$y = \int_0^y 1.\mathrm{d}y = \int_{-\infty}^x P(z)\mathrm{d}z = F(x) \tag{4.4}$$

$F(x)$ is the Gaussian cumulative distribution function (CDF). Thus given y from the random generator, we can obtain x via inverse CDF ($F^{-1}(y)$). However, the inverse CDF can be expensive to evaluate if a closed form analytical solution does not exist.

4.2.3.3 Box–Muller Approach

The Box–Muller Transform [9] is an efficient way to generate Gaussian random samples from uniform samples without dealing with inverse probability transforms. Let y_1 and y_2 be two uniformly distributed random variables over $[0, 1]$. Box–Muller generates a pair of Gaussian random numbers, x_1 and $x_2 \sim N_g(0, 1)$ according to

$$x_1 = \sqrt{-2\ln(y_1)} \times \cos(2\Pi y_2)$$
$$x_2 = \sqrt{-2\ln(y_1)} \times \sin(2\Pi y_2) \tag{4.5}$$

Solving for y_1 and y_2, one obtains

$$y_1 = e^{-\frac{\left(x_1^2 + x_2^2\right)}{2}} \tag{4.6}$$

$$y_2 = \frac{1}{\sqrt{2\Pi}} \tan^{-1}\left(\frac{x_2}{x_1}\right)$$

the Jacobian $\partial(y_1, y_2)/\partial(x_1, x_2)$ then agrees with (4.4).

$$\frac{\partial(y_1, y_2)}{\partial(x_1, x_2)} \propto P(x_1, x_2) \tag{4.7}$$

An alternative approach that is attributed to Marsaglia [10] is the polar form of Box–Muller. This approach is believed to be more efficient than the previous one; it avoids using the costly trigonometric functions. It is a form of "acceptance–rejection" sampling as it ends up discarding some sample points. The following equations summarise the polar method.

```
Pseudo Code: Polar Form of Box-Muller

while (r₂ >1) or (r₂ == 0)
{
        y₁= −1+2×rand()
        y₂= −1+2×rand()
        r₂= y₁×y₁ +y₂×y₂
}
return y₂* sqrt(-ln(r₂)/r₂)
```

Here, y_1 and y_2 are uniform random variables over the interval $[-1, 1]$. $\Pi/4$ or 78.5% of the time, the sample points fall within the unit circle, and 21.5% of the time, the sample points are rejected.

4.2.3.4 Ziggurat Method

"The ziggurat approach covers the target density with the union of a collection of sets from which it is easy to choose uniform points, then uses the rejection method." Marsaglia and Tsang [11].

The ziggurat method [11] is another efficient technique that relies on "acceptance–rejection" sampling. It avoids complex mathematical functions. However, it relies on tables and is more suitable for the generation of a large number of random sample points. Figure 4.1 illustrates a crude representation of the method; the graph resembles the ziggurat pyramids. In practice, the method divides the distribution into 255 equal area layers; each layer corresponds to a step of the pyramid in Fig. 4.1.

The method is summarised in the pseudo-code below. A random layer is selected. A random value x is selected in that layer. If x is smaller than x_i, i.e. in the region of the layer that completely falls underneath the Gaussian distribution in that layer, the sample point is accepted. If x is within the shaded region but falls in the critical region between x_i and x_{i+1}, the square with bolded dashes in the example of Fig. 4.1, it is possible that the sample point may fall above the distribution. Thus, another uniform sample point between $[y_i, y_{i+1}]$ is chosen to determine the corresponding "y" and the exact location of the sample point with respect to the bell curve. If $y < P(x)$, the sample point is accepted.

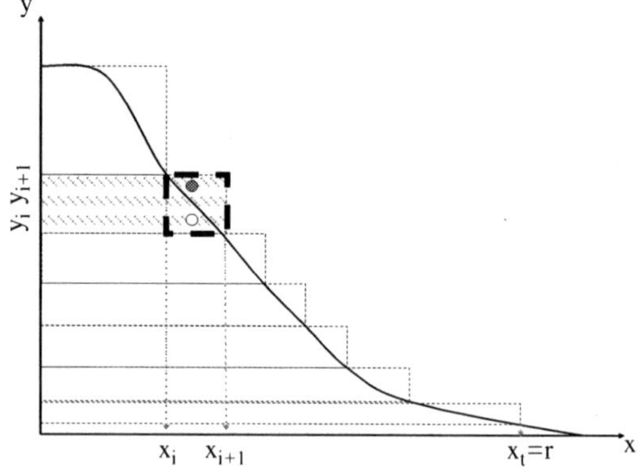

Fig. 4.1 Ziggurat method illustrated on right-side of a Guassian distribution. *The filled dot* is a rejected sample point. *The empty dot* is an accepted sample point. *The shaded layer* is the randomly selected ith layer

```
Pseudo Code: Ziggurat Method

Randomly pick layer i
x=rand()×x_{i+1}
        if (x < x_i)
        {return x}
        else if  (x_i < x < x_{i+1})
                { y=y_i +rand()*(y_{i+1}-y_i)
                 if  (y < P(x) )
                    return x}
```

If the lowest layer, layer t, is picked, the sample point falls in the tail layer. Recursive ziggurat can then be applied. Alternatively, "fall-back" algorithms for the tail region can be developed as is the case for the exponential and Gaussian distributions. The pseudo-code for a normal distribution is presented below.

```
Pseudo Code: Ziggurat Tail Sample Point Generation for Gaussian Distribution

Let the rightmost x_i=r.
u_1=rand()
u_2=rand()
do{
        x=-ln(u_1)/r
        y=-ln(u_2)
    } while 2y<x^2
return r+x
```

4.2.3.5 Generating Samples from Multi-Variate Gaussian Distributions

In two-dimensional space, the joint probability density function of two variables x and y, $P_{x,y}(x, y)$, satisfies the following relation [12].

$$P(x < X, y < Y) = \int_{-\infty}^{Y} \int_{-\infty}^{X} P_{x,y}(x, y)\mathrm{d}x\mathrm{d}y \qquad (4.8)$$

Two variables x and y are independent if the occurrence of value of one does not impact the other. For example, if the first coin is a tail, the second coin can still be head or tail with unchanged probability. Specifically,

$$P(x \cap y) = P(x) \cap P(y) \qquad (4.9)$$

The joint pdf of two independent variables can be derived from the individual pdfs.

$$P_{x,y}(x, y) = P_x(x)P_y(y) \qquad (4.10)$$

Hence, independent variables show no correlation, i.e. no relation or dependence between the random variables. Equation (4.12) illustrates the correlation coefficient

$\rho_{x,y}$ for two variables; $E[.]$ is the expectation function or the average function, and σ_x and σ_y are the standard deviations of x and y, respectively. By substituting (4.10) in (4.12), $\rho_{x,y}$ evaluates to zero for two independent variables.

$$E[x] = \int x.P_x(x)\mathrm{d}x \tag{4.11}$$

Two variables are correlated if $\rho_{x,y} \neq 0$.

$$\rho_{x,y} = \frac{E[XY] - E[X]E[Y]}{\sigma_x \sigma_y}$$
$$= \frac{\int\int xy \times P_{x,y}(x,y)\mathrm{d}x\mathrm{d}y - \int xP_x(x)\mathrm{d}x \int yP_y(y)\mathrm{d}y}{\sigma_x \sigma_y} \tag{4.12}$$

In multi-dimensional space, a set of independent Gaussian random variables $\{x_1, \ldots, x_d\}$ are distributed according to

$$P_{x_1,\ldots,x_d}(x_1, \ldots, x_d) = \prod_{i=1}^{d} \frac{1}{\sigma_i\sqrt{2\pi}} \exp\left(-\frac{(x_i - \mu_i)^2}{2\sigma_i^2}\right), \tag{4.13}$$

where $x_i \sim N_g(\mu_i, \sigma_i^2)$ ($i = 1$ to d in a d-dimensional space). Hence, it is common practice to generate the sample points for each random variable x_i independently. More complex techniques are needed for the case of correlated random variables. This will be discussed in the following section.

4.2.3.6 Correlated Gaussian Random Variables

In two-dimensional space, the joint pdf of the correlated Gaussian random variable x_1 and x_2, $\rho_{x1,x2} = \rho$, is

$$P_{x_1,x_2}(x_1, x_2) =$$

$$\frac{1}{2\pi\sigma_{x_1}\sigma_{x_2}\sqrt{(1-\rho^2)}} \exp\left(-\frac{\left(\frac{(x_1-\mu_{x_1})^2}{\sigma_{x_1}^2} + \frac{(x_2-\mu_{x_2})^2}{\sigma_{x_2}^2} - \frac{2\rho(x_1-\mu_{x_1})(x_2-\mu_{x_2})}{\sigma_{x_1}\sigma_{x_2}}\right)}{2(1-\rho^2)}\right) \tag{4.14}$$

In multi-dimensional space,

$$P_{x_1,\ldots,x_d}(x_1, \ldots, x_d) = \frac{1}{(2\pi)^{d/2}|\Sigma|^{1/2}} \exp\left(-\frac{1}{2}(\mathbf{x} - \boldsymbol{\mu})^T \Sigma^{-1}(\mathbf{x} - \boldsymbol{\mu})\right) \tag{4.15}$$

describes a correlated multi-variate Gaussian distribution. $\mathbf{x}$ is the random vector $[x_1, .., x_n]$. Σ is the covariance matrix. For simplicity, $\mathbf{x}$ is taken to be $\sim N_g(\mu, \Sigma)$.

$$\Sigma_{i,j} = \text{cov}(x_i, x_j) = E[x_i x_j] - E[x_i]E[x_j] \tag{4.16}$$

and

$$\Sigma_{i,i} = \text{cov}(x_i, x_i) \tag{4.17}$$

To generate correlated sample points from $N_g(\mu, \Sigma)$, a Cholesky decomposition of Σ is needed [13]. This is feasible because the covariance matrix is symmetric and positive semi-definite i.e., it satisfies (4.18).

$$\Sigma_{i,j} = \Sigma_{j,i} \quad \text{and} \quad z^T \Sigma z > 0 \tag{4.18}$$

$\forall$ non-zero vectors $\mathbf{z} \in R^d$. $\mathbf{z}^T$ is the transpose of $\mathbf{z}$.

A Cholesky decomposition finds a strictly positive lower triangle matrix M such that $\Sigma = MM^*$. M^* being the conjugate transpose of $\mathbf{M}$. The multiplication of $\mathbf{M}$ with independent Gaussian random vectors results in the desired correlated sample points. The pseudo-code below summarises the steps needed for sample point generation.

Pseudo_code: Correlated Gaussian Sample Point

```
Find M | MM*=Σ using Cholesky Decomposition
Z = randn() //uncorrelated, zero centered ~N g(0, I) Gaussian random vector
//in d-dimensions
return X= μ +M.Z
```

4.3 Monte Carlo Methods

"... theoreticians deduce conclusions from postulates, whereas experimentalists infer conclusions from observations. It is the difference between deduction and induction"
Hammersley and Handscomb [14]

From operations research to nuclear physics, Monte Carlo has been the science that experiments with random numbers.

One of the ancient applications to Monte Carlo that goes back to biblical days, more than 2000 years ago, is computing Π. Figure 4.2 illustrates a circle inscribed within a square. The area of the square is $4r^2$ and the area of the circle is $4\Pi r^2$. The ratio of the number of sample points that fall inside the circle to the ratio of sample points that are in the square is proportional to the ratio of the two areas and hence to Π.

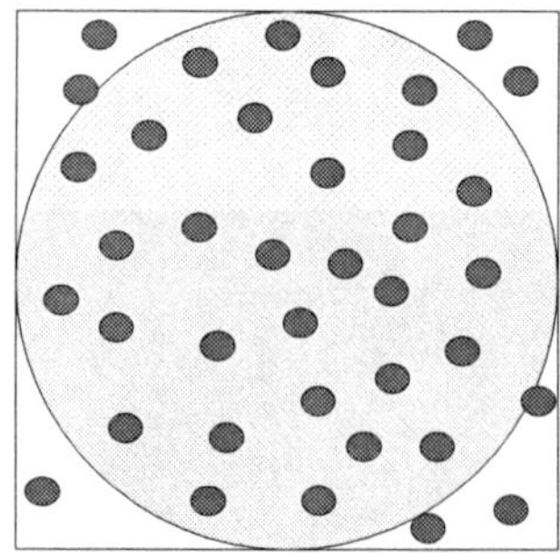

Fig. 4.2 The ratio of the dots inside the circle to the ratio of the dots inside the square (*all dots*) provides a way to compute Π

4.3.1 The Beginning

The year 1945 witnessed the first electronic computer and the birth of "Monte Carlo" [15]. What used to be known to the "old guard" as statistical sampling was for the first-time labelled by Von Neumann as "Monte Carlo" during the work on the atomic bomb in World War II. In 1947, Von Neumann wrote his famous letter providing a detailed outline for a statistical approach to solve the problem of neutron diffusion in fissionable material. The method was then code-named after Stan Ulam's Uncle who would borrow money from relatives and gamble in Monte Carlo. The methodology proved to be efficient at solving the problem compared to methods of differential equations. The scientists at Los Alamos Labs did not stop there. The Monte Carlo method proved to have lots of potential. Fermi, Von Neumann and Ulam would advocate the usage of the method to further solve deterministic problems.

Nowadays, the methodology has a wide application scope ranging from gaming, to finance, telecommunications and other fields. In the following sections, we will provide an overview of Monte Carlo for the related fields of numerical integration, statistical inference and reliability engineering.

4.3.2 Numerical Integration

4.3.2.1 Deterministic Methods

Numerical integration [16], or quadrature, is a large family of algorithms aimed at numerically evaluating definite integrals of the form (4.19). The need for numerical integration varies from the fact that a closed form for the integrand is not available, or is expensive to compute, to the fact that $f(x)$ is not fully available.

$$I = \int_a^b f(x)\mathrm{d}x \tag{4.19}$$

Fig. 4.3 Numerically solve for definite integrals of complex functions by dividing the interval into smaller intervals and relying on function approximations

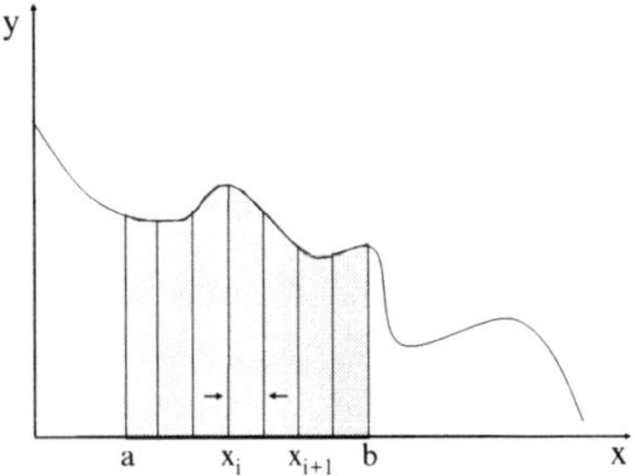

Solving for (4.19) is equivalent to solving the system of equations (4.20). Figure 4.3 illustrates an arbitrary function $f(x)$. The function is evaluated at N equally spaced points x_i over the interval $[a, b]$. The integral is then approximated as the sum of simpler integrals, often referred to as Riemann sum, that are easily evaluated over subintervals $[x_i, x_{i+k}]$; $i = \{1, 1 + k, 1 + 2k, \ldots, N - k\}$, k can take any value $\{1, 2, \ldots, N - 1\}$, and $x_1 = a$, $x_N = b$. The value chosen for k determines the complexity of the underlying method.

$$y(a) = 0; y(b) = I$$

$$\frac{dy}{dx} = f(x)$$

$$I \equiv \sum_i \int_{x_i}^{x_{i+k}} f(x)\mathrm{d}x \tag{4.20}$$

For arbitrary k, the method relies on Lagrange interpolating polynomials to simplify the function over subintervals. The extended midpoint rule [16, 17] is a special form where the integral over the interval $[x_i, x_{i+1}]$ is evaluated based on the midpoint $x_{i+1/2}$.

```
Pseudo_Code: Midpoint Rule

Pick a value for N
h=(b-a)/(N-1)
I=0
for (i=1; i<N; i++)
     { x_{i+1/2}=a+(i-1/2)h
       f_i=f(x_{i+1/2})
       I = I+h.f_i }
```

For $k = 1$, we obtain the Trapezoidal rule,

$$\int_{x_i}^{x_{i+1}} f(x)\mathrm{d}x \approx \frac{h}{2}[f(x_i) + f(x_{i+1})] \tag{4.21}$$

and for $k = 2$, we obtain Simpson's 1/3 rule.

$$\int_{x_i}^{x_{i+2}} f(x)\mathrm{d}x \approx \frac{h}{3}\left[f(x_i) + 4f(x_{i+1}) + f(x_{i+2})\right] \qquad (4.22)$$

More complex forms exist, e.g. using Simpson's 3/8 rule for $k = 3$, and other composite rules.

4.3.2.2 Curse of Dimensionality and Need for Monte Carlo Integration

Numerical methods like the trapezoidal rule in a Riemann's Sum fashion (equally spaced mesh) rely on a grid to partition the function over the region of integration. The number of grid points needed to obtain accurate estimates increases exponentially with dimensionality of the problem. Fig. 4 illustrates a $3\times3\times3$ grid in three-dimensional space. Partitioning the integration interval into intervals can become expensive and even prohibitive in high-dimensional problems. For example, to evaluate a six-dimensional integral using nine partitions (ten points) per dimension, the grid will require 10^6 evaluations of the function. If the analytical function is available, that may be accommodated; however, if the function requires excessive computations for evaluating each grid point, this can become expensive. In general, such methods are usually adopted for lower dimensional problems ($d < 5$); d is the number of dimensions.

To overcome this curse of dimensionality, Monte Carlo methods offer a more efficient way of computing the integrals with the same number of function evaluations. Monte Carlo spreads out the sample points uniformly over the integrand space; hence, no grid is used. More sophisticated methods include stratified sampling and Latin hypercube sampling [18] [19]. Such methods guarantee better spread of the sample points over the integrand space. In the following section, we will describe Latin hypercube sampling.

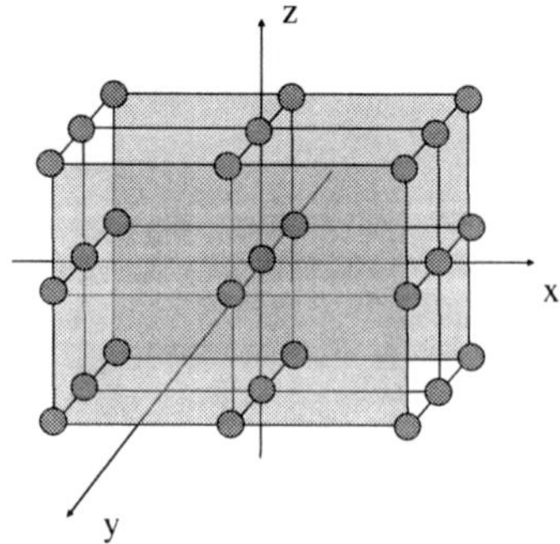

Fig. 4.4 Number of grid points increases exponentially with the dimensionality of the problem. An illustration in three-dimensional space of a $3 \times 3 \times 3$ grid

4.3.2.3 Latin Hypercube Sampling

By definition, a Latin square is a sparse grid square that has a sample point in each row and column. A Latin hypercube is a generalisation to multi-dimensional space. As illustrated in Fig. 4.5, Latin hypercube sampling spreads out the sample points, by dividing the space into a grid. It guarantees good sample point spread by excluding sampling in rows and columns that are covered in the preceding sample points. Hence unlike traditional grid sampling methods, the total number of sample points is arbitrary and determines the grid size differently from the traditional grid-based methods. Given this property, the method may not be good for capturing interactions [16]. Usually, it is most handy when the number of sample points is limited, for example due to function evaluations being very expensive.

Note that the implementation in Fig. 4.5 is just for conceptual understanding and efficient implementations can be sought in [16]. Finally, a more sophisticated form of sampling is the orthogonal sampling [16] that tries to generate the sample points all at once and tries to guarantee that the sample points are evenly spread and hence is a better representative of the sampling space.

> **Psuedo_code: Latin Hypercube Sampling**
>
> Generate M sparse sample points in d-Dimensional space
>
> 1. Divide each dimension into M segments resulting in M^d blocks.
> 2. Randomly pick a block from the available blocks.
> 3. Within the block pick a random sample point.
> 4. Eliminate the rows and columns of this block.
> Repeat, go to 2.

4.3.3 Statistical Inference

The act of inferring information from random samples to generalise the behavior of the population is known as statistical inference [20]. This information can include properties like the mean of the population, frequency probability and the assumed confidence in such estimates. Integration is fundamental to statistical inference. In fact, evaluating means, variances and errors all involve some form of integration. Monte Carlo integration plays a significant role in statistical inference.

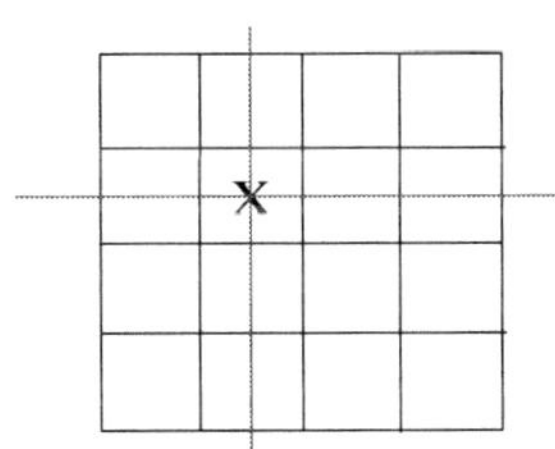

Fig. 4.5 Illustration of a Latin hypercube sample point and elimination of the corresponding rows and columns

4.3.3.1 Mean and Variance

Monte Carlo often deals with estimating the value of a certain property of a distribution called the estimand, given a certain set of random observations $\{y_1, y_2, \ldots, y_N\}$: the sample. Without loss of generality, let the sample size be N. The sample is merely a small set of the hypothetical larger set represented by the population. Box Hunter and Hunter [20] define the mean of the sample, $\bar{y}$, as "a measure of location" of the sample, and the variance of the sample, s^2, as a "measure of the spread" of the sample.

Let us say that the true population mean is μ and the true population variance is σ^2. The sample mean and variance are estimated according to (4.23) and (4.24).

$$\bar{y} = \sum_{i=1}^{N} \frac{y_i}{N} \tag{4.23}$$

$$s^2 = \sum_{i=1}^{N} \frac{(y_i - \bar{y})^2}{(N-1)} \tag{4.24}$$

Here, $v = (N-1)$ is the number of degrees of freedom; this is governed by the fact that the population mean is being replaced by an approximate average, hence the loss of one degree of freedom. The terms $(y_i - \bar{y})$ are often referred to as the residuals. In Table 4.1, the random variable y is assumed to be uniformly distributed over the range [0, 1]. A set of ten random sample points is generated according to the algorithms of Sect. 4.2. The sample is then replicated five times. The table summarises the estimated mean and variance for each of the samples. It also summarises the mean and variance of the estimated mean $\bar{y}$. Notice how the spread, or variance of $\bar{y}$, is smaller than the variance of y.

Table 4.1 y is a uniformly distributed random variable over the interval [0, 1]

1	0.08	0.23	0.79	0.11	0.20
2	0.04	0.71	0.04	0.57	0.70
3	0.46	0.90	0.19	0.71	0.72
4	0.03	0.40	0.49	0.57	0.74
5	0.22	0.32	0.79	0.70	0.79
6	0.76	0.44	0.88	0.20	0.33
7	0.73	0.98	0.41	0.51	0.66
8	0.18	0.65	0.99	0.08	0.85
9	0.52	0.26	0.12	0.06	0.87
10	0.61	0.74	0.98	0.89	0.99
Mean (y)	0.36	0.56	0.57	0.44	0.69
Var(y)	0.08	0.07	0.13	0.09	0.06
Mean ($\bar{y}$)	0.52				
Var ($\bar{y}$)	0.01				

Five sets of sample points are obtained with each sample size being 10. The mean and variance are summarised

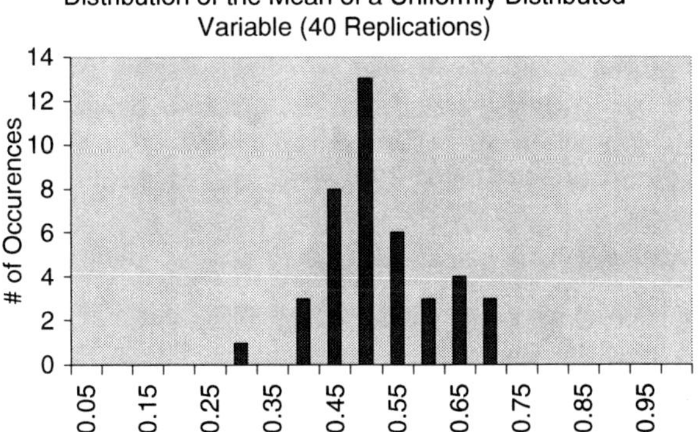

Fig. 4.6 The spread of the mean $\bar{y}$. $\bar{y}$ is the average of 10 sample points. A total of 40 replications are used to estimate the $\bar{y}$ distribution

Figure 4.6 illustrates the distribution for 40 replicas. This is often referred to by statisticians as the "Central limit tendency for averages." With the average being a sum of identically distributed independent random variables, its distribution tends to a Gaussian distribution, regardless of the underlying distribution of the random variable, y in this case. In general, $\bar{y}$ tends to a normal distribution as defined in (4.25) for large N, where N is the sample size. If y is normal, then (4.25) holds as a strict equality.

$$\bar{y} \sim N_g\left(\mu, \frac{\sigma^2}{N}\right) \tag{4.25}$$

The distribution of the sample variance is often skewed. If s^2 has v degrees of freedom, then $\chi^2 = vs^2/\sigma^2$ is a chi-squared distribution when y is normally distributed, where

$$E[s^2] = \sigma^2 \tag{4.26}$$

$$\mathrm{var}(s^2) = 2\sigma^4/v \tag{4.27}$$

When the distribution of y is not normal

$$\mathrm{var}(s^2) = \frac{2\sigma^4}{v}\left(1 + \frac{v}{2N}\gamma_2\right), \tag{4.28}$$

where

$$\gamma_2 = \frac{\mu_4}{\sigma^4} - 3 \tag{4.29}$$

and μ_4 is the fourth moment of the population distribution of y [20].

4.3.3.2 Confidence Intervals

Often the size of the sample is limited, and the estimate will have some error compared to the true population statistic. The question is whether one can determine the error of estimate and whether we can compute "an interval within which the true value almost certainly lies" [20]. This interval is often referred to as the *confidence interval*.

The confidence interval is associated with

- a confidence level $(100 \times (1 - 2\alpha))$ %, say 95%, and
- a confidence limit (the upper and lower bounds): [UB, LB].

For a given confidence level, the symmetric confidence interval for the mean estimate is determined by solving for Δ in (4.30).

$$
\begin{aligned}
\text{LB} &= \bar{y} - \Delta \\
\text{UB} &= \bar{y} + \Delta \\
\Pr(\mu &\in [\bar{y} - \Delta, \bar{y} + \Delta]) \cong (1 - 2\alpha)
\end{aligned}
\tag{4.30}
$$

Δ is therefore dependent on the distribution of $\bar{y}$. Assuming a normal distribution approximation of $\bar{y}$ based on the central limit theorem and following (4.25), the 95% confidence interval Δ can be computed as function of the standard deviation of the estimate $\bar{y}$ according to (4.31); $\text{Var}(\bar{y}) = s^2/N$ following (4.25), N is the sample size and $\phi(z)$ is the cumulative distribution function of a Gaussian normal distribution. The 95% confidence interval for $\bar{y}$ is, therefore, found to be equal to $[\bar{y} - 1.96s/\sqrt{N}, \ \bar{y} + 1.96s/\sqrt{N}]$. Note that in (4.31), z is a standard normal random variable $\sim N_\text{g}(0, 1)$.

$$
\begin{aligned}
\Delta &= cs/\sqrt{N} \\
\Pr\left(\bar{y} - cs/\sqrt{N} < \mu < \bar{y} + cs/\sqrt{N}\right) &\cong (1 - 2\alpha) \\
\Pr(-c < z < c) &= (0.95) \\
\Pr(z < c) &= (1 - \alpha) = 0.975 \\
c &= \varphi^{-1}(0.975) = 1.96
\end{aligned}
\tag{4.31}
$$

Figure 4.7 illustrates the 99% confidence intervals for the means estimated in Fig. 4.6. We notice that the true population mean ($\mu = 0.5$) often falls within the computed intervals. In general, it is desired that the confidence interval range be small compared to the estimate itself; i.e., we want the variance in the estimate, often referred to as the error in the estimate, to be small compared to the estimate. This way, the confidence interval is relatively small and the sample estimate is a good representative of the true mean.

In cases where the confidence in the amount of variation of the estimate is critical to the study, confidence limits for the variance σ^2 are desired. Well-known

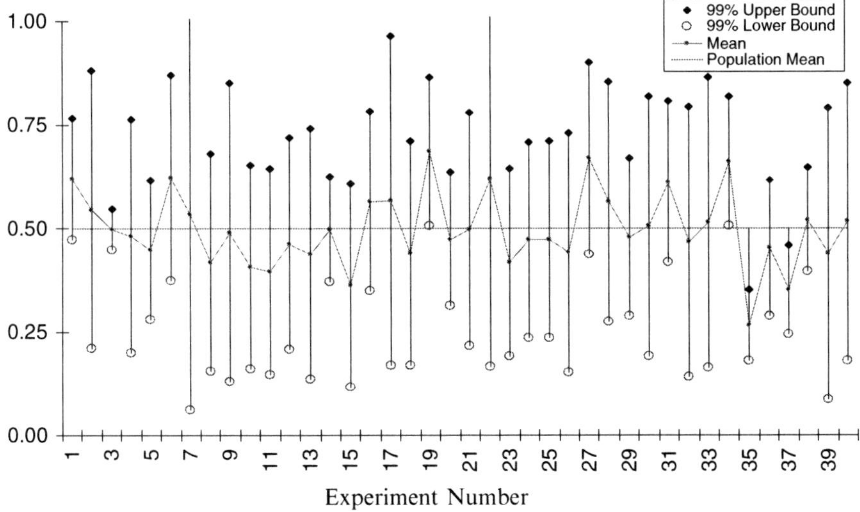

Fig. 4.7 Ninety-nine percent confidence interval for the samples in Fig. 4.6 vs. the true population mean

closed forms are available for a normally distributed population according to (4.32).

$$\Pr\left(\frac{vs^2}{B} < \sigma^2 < \frac{vs^2}{A}\right) = (1 - 2\alpha), \tag{4.32}$$

where A and B are the values for the χ_v^2 distribution for probabilities (α) and $(1 - \alpha)$.

4.3.3.3 Statistical Inference for Proportions (Probabilities)

A Bernoulli trial is an event whose outcome takes one of two values {"pass" or "fail"} or {"0" or "1"}. An example of a Bernoulli trial can be a coin toss; for example, we can assume success if we obtain a head. Or we assume success, if we obtain a number larger than 4 when throwing a die, or a sum larger than 10 when throwing a pair of dice. Note that the probability of success is not necessarily equal to the probability of fail.

Often the Bernoulli trial is represented by a random variable that is modeled by the indicator function, I, where

$$
\begin{aligned}
I(x) &= \begin{cases} 0 & x = \text{fail} \\ 1 & x = \text{pass} \end{cases} \\
E[I] &= 0 \times (1 - p) + 1 \times p = p \\
\text{var}(I) &= p - p^2 = p(1 - p) = pq
\end{aligned}
\tag{4.33}
$$

Now assume that we repeat the event several times, say n times; this is often referred to as the number of trials. The number of successes or the proportion of time we obtain a success is distributed according to the binomial distribution. To explain this further, we will consider the example of throwing a die. The throw is successful anytime we obtain a number greater than 4. Thus the probability of success $p = 2/6$, and the probability of fail $q = 4/6$. Note that $p + q = 1$. Let the number of trials be three. We expect the probability of the number of successful outcomes, $P(y)$, to be as follows:

- "0" successes with probability $P(y = 0) = q^3 = (4/6)^3$;
- "1" success with probability $P(y = 1) = {}^3C_1 \times p.q^2$. ${}^nC_m$, the combination function, accounts for the order of occurrence. For example, it is possible for the successful throw to happen during the first or the second or the third time;
- "2" successes with probability $P(y = 2) = {}^3C_2 \times p^2 q$;
- "3" successes with probability $P(y = 3) = p^3$.

In general, given n trials, $P(y = m)$ is distributed according to the binomial distribution

$$\text{binomial}(n, p, y) = P(y) = {}^nC_m.p^y q^{n-y}.$$

$$ {}^nC_m = \frac{n!}{y!(n - y)!} \tag{4.34}$$

$n!$ in the combination definition is the factorial of "n" and is equal to $n \times (n - 1) \times (n - 2) \times \ldots \times 2 \times 1$. For example, $3! = 3 \times 2 \times 1 = 6$.

Figure 4.8 summarises an experiment where the number of trials $n = 30$. One can derive from the binomial distribution, the probability of a success greater than a

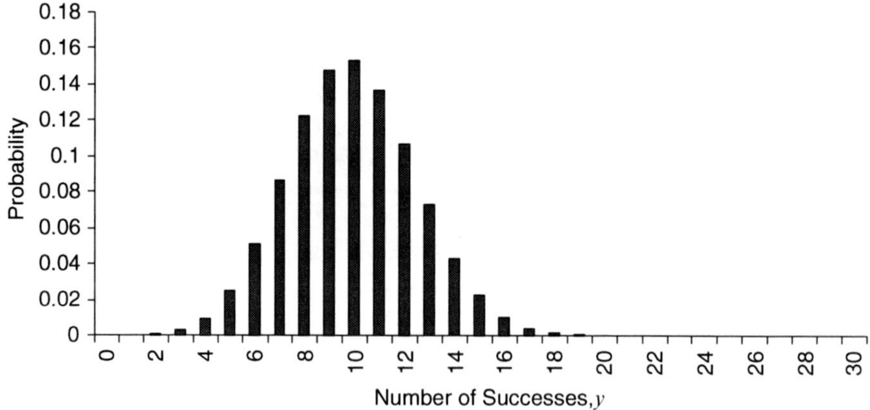

Fig. 4.8 A die throw is considered successful when we obtain a number larger than 4 ($p = 2/6$). The number of trials is 30. $P(y)$ is the probability of obtaining y successful trials

given number, for example, $P(y > 20)$. Also, we can find the expected number of successes and the variation in the expected number of successes; this is often referred to as the mean and variance of the binomial distribution.

The mean and variance of the binomial distribution can be derived as follows:

$$E[y] = \sum_{y=0}^{n} y.P(y) = \sum_{y=0}^{n} y.\text{binomial}(n, p, y)$$

$$\text{var}(y) = \sum_{y=0}^{n} y^2 P(y) - E(y)^2$$

$$(4.35)$$

Alternatively, we can consider that the binomial distribution represents the outcome of many independent Bernoulli experiments. Hence,

$$y = I_1 + I_2 + ... + I_n \tag{4.36}$$

and

$$E[y] = E[I_1] + E[I_2] + ... + E[I_n] = np$$

$$\text{var}(y) = \text{var}(I_1 + I_2 + ... + I_n) = n\text{var}(I) = npq$$

$$(4.37)$$

For the example in Fig. 4.8, the expected number of successes $E[y] = 30 \times 2/6 = 10$.

Therefore, on average we expect to obtain ten successes out of 30 throws. The variance $\text{var}(y) = npq = 30 \times 2/6 \times 4/6 = 6.6$, and the error or standard deviation in the expected value is 2.58. Although these are not whole numbers, they provide us a good idea about the expected number of successes or a most probable range for the number of successful attempts.

As the number of trials, n, increases, the binomial distribution can be approximated by a normal distribution. This enables significant compute time savings. For the approximation to be good, it is recommended that $np \geq 10$ and $n(1 - p) \geq 10$ [20].

$$P(y > y_0) = P\left(z = \frac{y - np}{\sqrt{npq}} > \frac{y_0 - np}{\sqrt{npq}}\right), \tag{4.38}$$

where z is $\sim N_g(0, 1)$. Yates introduced an adjustment to the y_0 value by 0.5 [16]. For example, we can approximate $P(y < 8)$ by $P(y < 8.5)$ to account for the discontinuity in the binomial distribution, resulting from the integral domain of the binomial random variable.

4.3.3.4 Statistical Inference for Frequencies

As n increases and p decreases, the binomial distribution tends to a Poisson distribution [20]. For the approximation to be valid, it is recommended that

$n > 100$ and $np \leq 10$. In general, a Poisson distribution has been linked to the rate or frequency of occurrence. For example, the probability of back injury for an employee at a given job is very small, say 1×10^{-4}/h. The employee, however, works full time and thus spends around 2,000 h per year. Hence, the number of back injuries for him in a given year is distributed according to a binomial distribution $p = 1 \times 10^{-4}$ and $n = 2,000$. From this, we compute the average rate of injuries per given year to be $\lambda = np = 0.2$. The probability of y injuries per year can be computed according to the Poisson distribution (4.39).

$$P(y) = \frac{e^{-\lambda}\lambda^{y}}{y!} \tag{4.39}$$

The mean and variance of a Poisson distribution are both equal to λ.

$$\begin{aligned} E[y] &= np = \lambda \\ \mathrm{var}(y) &= \lambda \end{aligned} \tag{4.40}$$

For sufficiently large λ, the Gaussian $N_{\mathrm{g}}(\lambda, \lambda)$ can be used as an approximation to the Poisson distribution. Figure 4.9 illustrates Poisson probability $P(y)$ for $\lambda = 1.5$ and $\lambda = 10$. We note that the probability density for $\lambda = 10$ starts to resemble the Gaussian bell curve.

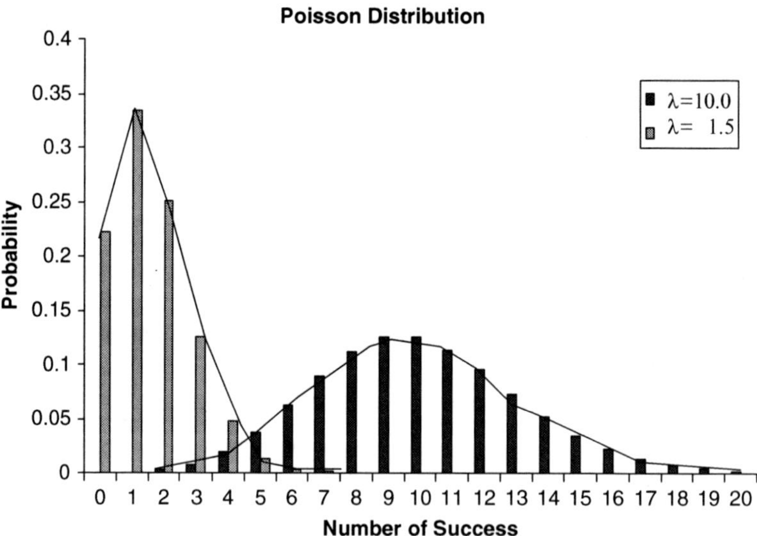

Fig. 4.9 Poisson distribution tends to a Gaussian for large λ

4.3.3.5 Reliability Engineering and Monte Carlo: Acceptance Space Complexity

The science of reliability engineering, or design yield analysis, covers a wide range of engineering domains from mechanical to civil engineering and from telecommunications to robotics [21]. The goal is always the same, which is to guarantee that a system operates with high yield and has a very low fail/error probability. Often for communications engineering systems, designers are more worried about the bit error rate and false positives. In semiconductor engineering, designers are worried about chip yield. When the design consists of many parts with a given fail probability, statistical inference techniques need to be employed.

To generalise the problem, let us assume that the design reliability is measured by a set of pre-defined functions denoted $f_j(\mathbf{x})$. For example, $f_j(\mathbf{x})$ can be performance metrics of the designed circuit, where $\mathbf{x} = \{x_1, x_2, \ldots, x_d\}$ denotes the set of independent variables in the d-dimensional design parameter space. We will assume, henceforth, that x_i, $i \in \{1,.., d\}$, are independent normal Gaussian random variables; an example is that x_i can represent the threshold voltage of the design transistors.

Let $\mathbf{f}^0$ be the set of corresponding critical values that determine the acceptability region for the design

$$\text{pass} \Leftrightarrow f_j(\mathbf{x}) < f_j^0; \quad \mathbf{x} \in R^d, \tag{4.41}$$

where $j = \{1, \ldots, m\}$; m is the number of performance metrics. As is the case with most circuit designs, $f_j(\mathbf{x})$ is usually obtained via simulation, as opposed to closed form approximations because simulation is more accurate.

It is possible to estimate the design yield by integrating the pdf of the performance metrics over the simple acceptability region (see Fig. 4.10a) according to integral I_1 in (4.42).

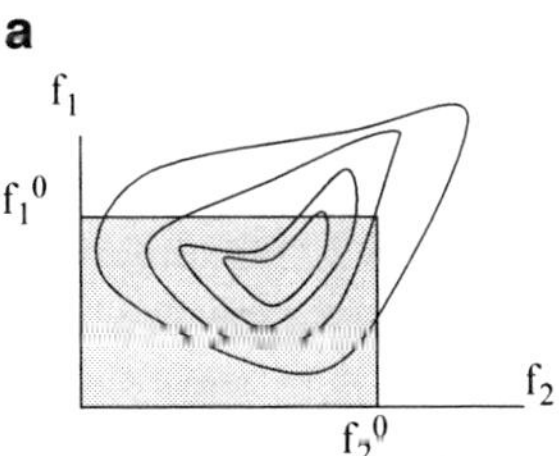

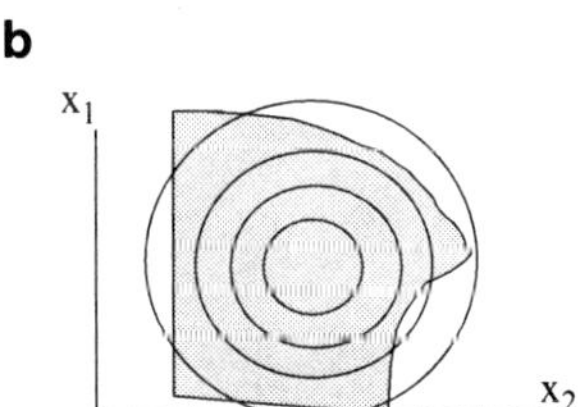

Fig. 4.10 (a) Estimating yield in design metric space. (b) Estimating yield in the variability space

However, often the true pdf is not known and it is only possible to obtain first-order approximations. This is not desirable, especially that our goal is to estimate low fail probabilities.

It is possible to avoid such approximations of the performance space distributions, by integrating the pdf of the sources of variability (x_i) over the usually complex, acceptability region A (see Fig. 4.10b), as indicated in integral I_2 in (4.42) [21].

$$\text{Yield} = I_1 = I_2$$

$$I_1 = \int_0^{\mathbf{f}_0} \text{pdf}(\mathbf{f}(\mathbf{x})) d f$$

$$I_2 = \int_A \text{pdf}(\mathbf{x}) d x \tag{4.42}$$

Because the boundary in I_2 (4.42) is complex, we resort to statistical methods, mainly Monte Carlo sampling methods, to solve such integrals.

In fact in simple Monte Carlo, each sample point is selected according to the design paramater distributions. The design is then evaluated using the selected sample points. Like a Bernoulli trial, the design will fail or pass with a given probability. The only difference is that we do not know what p is apriori. So we rely on multiple sample points to estimate the probability of success, or the corresponding failure probability P_f. P_f and its variance σ_{Pf}^2 may be estimated from N random sample points that are independently drawn according to $P(x)$. Each sample point, $\mathbf{x}^k$, $k = \{1, \ldots, N\}$, will be associated with a given indicator function $I(\mathbf{x})$.

$$I(\mathbf{x}) = \begin{cases} 0, \text{pass}(\mathbf{f}(\mathbf{x}) < \mathbf{f}^0; & \mathbf{x} \in R^d) \\ 1, \text{fail}(\mathbf{f}(\mathbf{x}) > \mathbf{f}^0; & \mathbf{x} \in R^d) \end{cases} \tag{4.43}$$

$$P_f = \frac{1}{N} \sum_{k=1}^N I(\mathbf{x}^k), \quad \text{and} \quad \sigma_{Pf}^2 = \frac{P_f(1 - P_f)}{N} \tag{4.44}$$

Figure 4.11 illustrates a simple Monte Carlo example where $\mathbf{f}(\mathbf{x}) = \mathbf{x}$ in one-dimensional space (a single design parameter). The example assumes that $\mathbf{x}$ is uniformly distributed over [0, 1]. We note that the estimate approaches the true probability as the number of sample points increases. In fact according to (4.44), the error in the estimate is proportional to $1/\sqrt{N}$; N is the number of sample points.

4.3.3.6 Resampling: Cross-Validation and Bootstrapping

Bootstrapping, jackknifing and cross-validation [22] are resampling methods used to estimate the precision of certain statistics (mean, variance and probabilities).

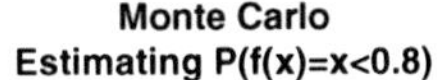

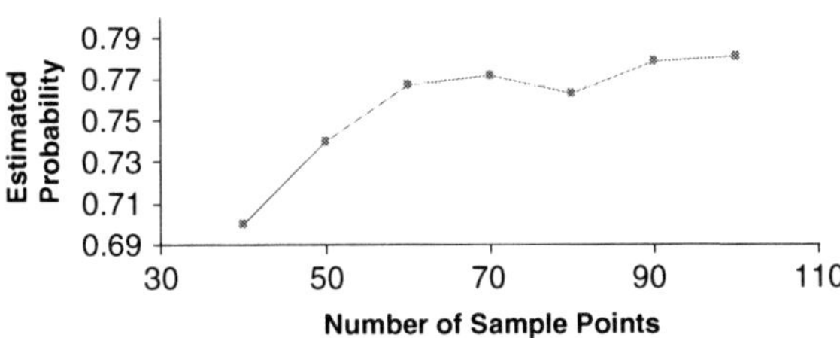

Fig. 4.11 Monte Carlo estimation. Estimate converges to the true probability as the number of sample points increases

Particularly for situations where there is no knowledge of any underlying sampling distribution, or if evaluating the sample points is very expensive, bootstrapping/resampling is used for further inference from the set of available sample points. Rather than relying on parametric assumptions of the underlying distribution, these techniques rely on resampling to study the variability in the statistic.

- Bootstrapping performs random sampling with replacement from the subset under study. Hence bootstrap is Monte Carlo on a set of pre-sampled sample points.
- Jackknifing leaves out a sample point from the sampling subset,
- Cross-validation leaves out a subset of the data; the rest of the data is used to build the model which is then validated on the remaining subset.

Figure 4.12 illustrates an experiment where bootstrapping is used to estimate the probability of a head in a coin toss. Ten trials are performed and the outcome sample $s = $ [H T H H T H T H T H] is saved. The original sample indicates that the probability is "0.6." This sample is then re-sampled, or replicated, 100 times by bootstrapping, i.e. randomly selecting ten values each with replacement. Each sample replication, r_i, $i = \{1,..., 100\}$, is then used to estimate the probability of a head. This is represented by the solid line in Fig. 4.12. The dashed line represents true Monte Carlo estimations if the 10-toss experiment is to be repeated 100 times. The bootstrap experiment gives a good feel for the expected range or spread of the estimated statistic.

Of course, the result of bootstrap replications is dependent on how well the original sample represents the true data. For example, if the only sample we have is an "all tails" sample, then definitely the following bootstrap replications will be misleading; however, so is the "all tail" original sample, the problem resides in the source of data. The bootstrap method is only aimed at maximizing the knowledge from whatever is available.

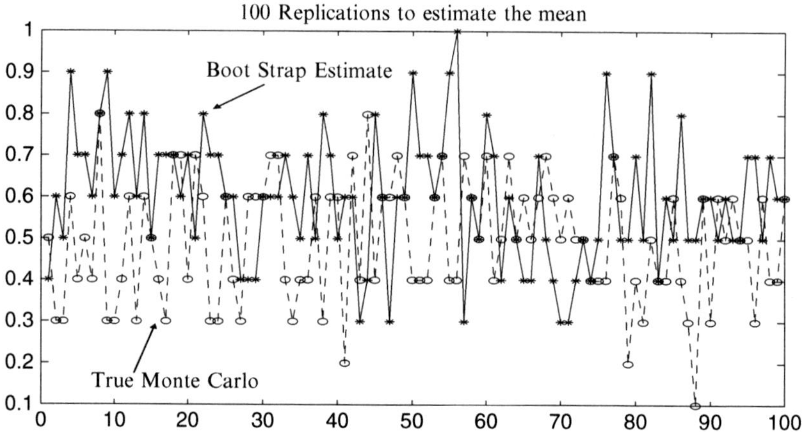

Fig. 4.12 Ten sample points are used to estimate the probability (0.5) of a head in an unbiased coin. One hundred replications are performed. The Monte Carlo estimate is plotted in dashes. Bootstrap estimate from one of the 10 toss samples is plotted in solid line. One hundred bootstrap "set-of-10" samples are derived from a sample whose estimate was 0.6

4.3.4 Rare Event Estimation and Monte Carlo

We demonstrated in Fig. 4.11 a converging estimate trend as the number of sample points increases. This is inherent to (4.44) as the standard deviation in the estimate is proportional to the square root of the number of sample points. This dependence on the number of sample points becomes more evident as we study the confidence interval of the probability of fail estimate, P_f, and as we deal with very low fail probabilities.

As we discussed before, the true value of the estimate is expected to fall in the confidence interval, $[P_f - \Delta, P_f + \Delta]$, with a probability $(1 - 2\alpha) \times 100\%$, also referred to as the confidence level. Smaller confidence interval ranges along with higher confidence levels are preferred. Specifically, it is desired that confidence intervals be small relative to estimate, implying reduced error in the estimate. For this study, we assume that Δ can be related to the estimate by $\Delta = \gamma P_f$, and we require a small γ.

Similar to the previous sections, and assuming normal approximation for the estimate, we find the 95% confidence interval to be $[P_f - 1.96\sigma_{Pf}, P_f + 1.96\sigma_{Pf}]$. Then γ is related to σ_{Pf} according to (4.45).

$$\frac{1.96 \times \sigma_{P_f}}{P_f} = \gamma \tag{4.45}$$

From (4.44) and (4.45), one can then derive the number of sample points needed to obtain a 95% confidence interval whose range is within $-/+ \gamma$ of the estimate as shown in (4.46).

$$N = \frac{1.96^2}{\gamma^2} \cdot \frac{(1 - P_f)}{P_f} \tag{4.46}$$

For small P_f values, N is inversely proportional to P_f, and we expect standard Monte Carlo methods to become less efficient as P_f decreases. To demonstrate this in a simple example, we assumed that $f(x) = x$, where x is a random $\sim N_g(0, 1)$; i.e. x follows a standard normal distribution. We rely on Monte Carlo to estimate the probability of fail, $P_f = \text{Prob}(x > z_0 = f^0)$. Table 4.2 presents the values of N needed to estimate P_f accurately with $\gamma = 10\%$ according to (4.46). The number of sample points increases dramatically as z_0 increases (P_f decreases). The smaller probability values are often referred to as the tail probabilities (see Fig. 4.13); the corresponding z_0 values are far from the mean.

Also, Table 4.2 lists what would be the equivalent runtime if $f(x)$ involved SPICE-like simulations of a moderately large memory design. While the runtime can be further minimised by using fast simulators, or even response surface models, the number of sample points needed for estimating low failure probabilities is neither practical nor reasonable. It is obvious that there is a need for more efficient statistical methods that are capable of estimating such low fail probabilities with good confidence and reasonable resources.

Table 4.2 Number of Monte Carlo sample points needed to estimate the probability $P_f = \text{Prob}$ $(x > z_0)$ with a 95% confidence interval $- [P_f - 0.1P_f, P_f + 0.1P_f]$

z_0	Probability value (P_f)	Number of Monte Carlo sample points (N)	Approximate runtime
0	0.500	4.0e2	4.3 h
1	0.159	2.1e3	23.0 h
2	0.0228	1.7e4	7.8 days
3	0.00135	2.9e5	130 days (too long!)
4	3.169E-05	1.3e7	16 years (too long!)
5	2.871E-07	1.4e9	1736 years (too long!)

The corresponding simulation runtime in days, if a SPICE-like simulator is used

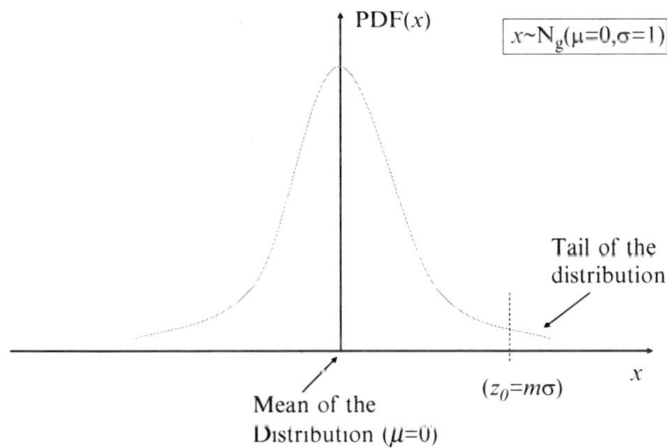

Fig. 4.13 Sketch of the tail of the distribution for a Gaussian variable. For a standard Gaussian variable, $\sigma = 1$

4.4 Variance Reduction and Importance Sampling

4.4.1 An Overview of Variance Reduction Methods

We saw in the previous section that the error in the estimate decreases as the number of sample points, N, increases in the standard Monte Carlo method. For very small probability estimates, the sample size is $\alpha 1/P_f$. Hence the accuracy of the estimate degrades for rare event estimation, unless the number of sample points increases dramatically. Variance reduction techniques are statistical sampling methods that enable relying on a reasonable number of sample points compared to typical Monte Carlo method. In fact, the term "variance reduction" arises from the ability of such methods to reduce the variance or the error of the estimate, given a fixed number of sample points.

Most common variation reduction methods include control variates, antithetic variates, quasi-random numbers, stratified sampling and importance sampling. In this section, we will briefly cover a few of these methods; we refer the reader to [23–25] for more details. Then we focus on importance sampling methods.

4.4.1.1 Control Variates

The basic concept relies on exploiting the correlation between random variables to reduce the variance of the estimate. An example of correlated variables would be traffic accidents and weather conditions.

The general problem can be formulated as follows. Let us assume that we want to estimate $P_m = E[x_1]$. Also, let us assume another variable x_2 that is correlated to x_1. While the typical approach solves for the expected value of $\hat{\theta}$ (4.47), the method of control variates solves for the expected value of $\hat{\theta}_c$ (4.48), and x_2 is assumed to be a control variate of x_1; c is a constant.

$$\hat{\theta} = x_1 \tag{4.47}$$

$$\hat{\theta}_c = x_1 + c(x_2 - E[x_2]) \tag{4.48}$$

Hence, we have $E[\hat{\theta}_c] = E[\hat{\theta}]$. To enable a lower variance for the estimate, the control variates method requires optimal c to be

$$c^* = -\frac{\mathrm{Cov}(x_1, x_2)}{\mathrm{Var}(x_2)} = \rho_{x_1 x_2} \mathrm{Var}(x_1) \tag{4.49}$$

In this scenario, $\mathrm{Var}(\hat{\theta}_c) < \mathrm{Var}(\hat{\theta})$ as long as the $\mathrm{Cov}(x_1, x_2) \neq 0$. The methodology though requires good knowledge of x_2; otherwise we assume that $\mathrm{Cov}(x_1, x_2)$, $\mathrm{Var}(x_2)$ and $E[x_2]$ can be easily generated by Monte Carlo experiments.

Equation (4.50) describes the more generic case of multiple control variates $(x_2, x_3, \ldots, x_n)$.

$$\hat{\theta}_c = x_1 + c_2(x_2 - E[x_2]) + \ldots + c_n(x_n - E[x_n]) \tag{4.50}$$

Again, it is clear that $E[\hat{\theta}_c] = E[\hat{\theta}]$. The optimal c_i can be obtained from Monte Carlo experiments as before. A good estimate for c_i would be to find the least squares solution (b_i) of the linear regression (4.51) and setting $c_i = -b_i$.

$$x_1 = b_2 x_2 + \ldots + b_n x_n + \varepsilon \tag{4.51}$$

ε is the regression error. Note that this is the reason why the technique is also known as regression sampling.

4.4.1.2   Antithetic Variates

Antithetic means being in the direct and opposite direction. Two sample points c_1 and c_2 are antithetic if $c_1 - c_2$. In general, given two variables x_1 and x_2 having the same distribution as x, then $E[x]$ can be derived according to (4.52).

$$E[x] = E\left[\hat{\theta} = \frac{x_1 + x_2}{2}\right] \tag{4.52}$$

$$\mathrm{Var}(\hat{\theta}) = \frac{\mathrm{Var}(x_1) + \mathrm{Var}(x_2) + 2\mathrm{Cov}(x_1, x_2)}{4} \tag{4.53}$$

If x_1 and x_2 are independent, then $\mathrm{Cov}(x_1, x_2) = 0$, and $\mathrm{Var}(x_1) = \mathrm{Var}(x_2)$. In that case, variance of $\hat{\theta}$ according to (4.54) is

$$\mathrm{Var}(\hat{\theta}) = \frac{\mathrm{Var}(x_1)}{2} \tag{4.54}$$

If x_1 and x_2 are negatively correlated, then $\mathrm{Cov}(x_1, x_2) < 0$ and $\mathrm{Var}(\hat{\theta})$ is smaller than (4.54). Example antithetic variates are $(x, 1 - x)$ for a uniform random variable over $[0, 1]$, and $(x, 2\mu - y)$ for a Gaussian random variable $N_g(\mu, \sigma^2)$.

4.4.1.3   Quasi-Monte Carlo Methods

As we discussed earlier, grid-based integration offers good and even coverage of the space at the expense of slow convergence rate that is proportional to $N^{-2/d}$ [25] (d being the dimensionality of the problem); the convergence rate suffers as the dimensionality of the problem increases. Typical Monte Carlo integration, on the contrary, employs random sample points and hence eliminates the dimensional dependency. It offers a good convergence rate that is proportional to $1/\sqrt{N}$ (4.44). However, the sample points are not guaranteed to provide good coverage of the sampled space (or region of integration). For example, sample points that are uniformly distributed over a cube are likely to be well spread but are not guaranteed to be regularly spread, as is the case of grid-based coverage of the cube.

Quasi-Monte Carlo Methods [16, 26] are sampling-based integration methods similar to typical Monte Carlo methods. However, they use sample points generated from special deterministic sequences, as opposed to traditional pseudo-random number generators. The deterministic sequences are designed to enable more evenly spread sample points that fill the d-dimensional cube more efficiently (in a uniform distribution manner). Quasi-Monte Carlo sequences offer more regular coverage of the d-dimensional cube, and the deterministic sequences are often referred to as low discrepancy sequences. This helps improve the convergence rate compared to traditional Monte Carlo. Quasi-Monte Carlo integration convergence rate is found to be proportional to $1/N^{1-\varepsilon}$, $\varepsilon > 0$, though ε hides some dependency on the dimensionality of the problem and often the convergence rate is found to be proportional to $(\ln N)^d/N$. To measure the discrepancy, or goodness of spread of samples, of a deterministic sequence, we rely on the star discrepancy metric [25].

The Star Discrepancy $D_N^*(x_1, \ldots, x_N)$: It is a measure of the "regularity of the sample." For specific sequences, there must be an accompanying bound on D_N^* [26].

Let P be a sample point set $= \{\mathbf{x_1}, \ldots, \mathbf{x}_N\}$ over the unit cube in d-dimension.

$$D_N^*(P) = \sup_{B \in J*} \left| \frac{A(B; P)}{N} - \lambda_s(B) \right|, \tag{4.55}$$

where J^* is the set of all d-dimensional rectangular subsets of the *unit cube*; each rectangle B will have one vertex at the origin, and the other vertices are arbitrary points in d-dimensional space. "sup" stands for supremum. λ_s stands for the Lebesgue measure (volume), and $A(B; P)$ stands for the number of sample points that lie inside B. To summarise, for each arbitrary rectangle with vertex at origin, we try to find the ratio of the number of sample points that lie in B, $A(B; P)$, to the total number of sample points. This ratio is then compared to the ratio of the volume of B to the volume of the unit cube. The supremum over all possible rectangles is the discrepancy of the set P.

Open versus closed sequences: Open sequences are characterised by incremental quality; i.e. given a set of N sample points, one can pick an additional sample point from the sequence while maintaining the low discrepancy property of the sequence. Examples of open sequences are Sobol' and Halton [16]. On the contrary, a closed sequence provides a new set of sample points for a desired number of sample points.

Tables 4.3 and 4.4 summarise the lower (upper)-bound discrepancy as a function of the number of sample points of Sobol', Halton and Monte Carlo distributions.

Discrepancy is based on the unit cube and is a measure of the maximum error due to the sample points estimating the volume of any rectangular region inside the unit cube. This is based on an algorithm proposed in [27].

Figure 4.14 compares the spread of the data in two-dimensional space for regular uniform sampling (Monte Carlo), vs. Sobol' sequence and Halton sequences.

Table 4.3 Discrepancy lower bound (*1e-3)

#Sample points	1,000	5,000	10,000
Sobol'	5.3	1.3	0.7
Halton	6.5	1.7	0.9
Rand	43	20	18.2

Table 4.4 Discrepancy upper bound (*1e-3)

#Samples	1,000	5,000	10,000
Sobol'	5.9	2.3	1.7
Halton	7.3	2.6	1.9
Rand	44	21	19

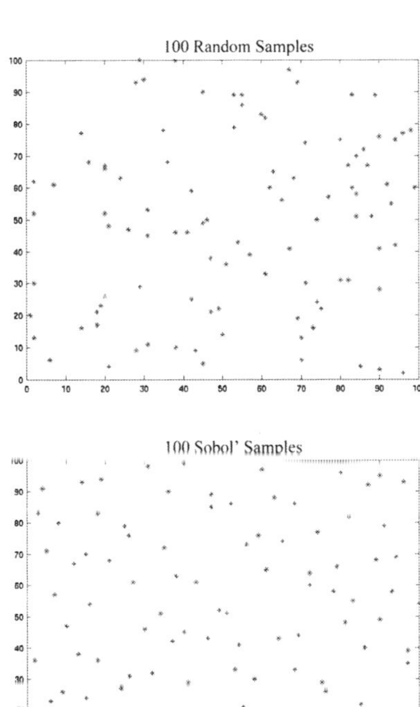

Fig. 4.14 Spread of 100 sample points in two-dimensional space (a) regular Monte Carlo and (b) Sobol' sequence

4.4.1.4 Stratified Sampling

As we discussed in the beginning of the chapter, stratified sampling is a well-known sampling technique that is based on dividing the population into strata [23]. A simple example to consider would be the case of a pot that has 100 balls, 50 of type "A" and 50 of type "B"; type "B" balls are heavier than type "A" balls. If we want to find the average weight of the balls with ten sample points, we can choose two approaches:

1. Select ten sample points randomly. In this scenario, one can obtain four "A" and six "B" type balls, or even ten "A" and 0 "B."
2. Divide balls into two strata "A" and "B." Select five "A" sample points and select five "B" sample points. In this scenario, the estimate is based on the proper "A"/"B" proportions, and $E[x] = \frac{1}{2} E[x|A] + \frac{1}{2} E[x|B]$ has smaller variance compared to scenario "1."

The reasoning behind this can be justified as follows. Let us compare the following: $\mathrm{Var}(x_A + x_B)$ and $\mathrm{Var}(x_1 + x_2)$. x_A and x_B are independent samples obtained from strata A and B, respectively, each of size N. x_1 and x_2 are two independent random samples of size N each.

Equation (4.56) and (4.57) describes the relation between the variance and conditional variance. Note that the implied statement is based on the fact that Var (.) and $E[\mathrm{Var}(.)]$ are positive quantities.

$$\mathrm{Var}(x) = E[\mathrm{Var}(x|.)] + \mathrm{Var}(E[x|.]) \Rightarrow \mathrm{Var}(x) \geq E[\mathrm{Var}(x|.)] \tag{4.56}$$

$$E[\mathrm{Var}(x|.)] = \frac{1}{m} \sum_{i=1}^{m} \mathrm{Var}(x|I_i)) \tag{4.57}$$

Given m strata: $I_1, \ldots, I_m$. From (4.57), we can derive the following:

$$\begin{aligned}
\mathrm{Var}(x) &\geq \frac{1}{2} \sum_{i=1}^{2} \mathrm{Var}(x|I_i)) = \frac{\mathrm{Var}(x_A) + \mathrm{Var}(x_B)}{2} = \frac{\mathrm{Var}(x_A + x_B)}{2} \\
\mathrm{Var}(x_1 + x_2) &= \mathrm{Var}(x_1) + \mathrm{Var}(x_2) = 2\mathrm{Var}(x) \\
\mathrm{Var}(x_1 + x_2) &\geq \mathrm{Var}(x_A + x_B)
\end{aligned} \tag{4.58}$$

4.4.2 Importance Sampling

> *Importance Sampling in Monte Carlo Simulation is the process of estimating something about a distribution using observations from a different distribution.* (T. Hesterberg)

By analysing the standard Monte Carlo simulation method, one realises that Monte Carlo works well; however, for rare event estimation of tail probabilities, it wastes a lot of time sampling around the mean rather than in the tails. Importance sampling [28] is a well-known variance reduction technique that gets around this problem by distorting the (natural) sampling distribution, $P(x)$, to produce a higher proportion of sample points in the important region(s) (see Fig. 4.15). Mathematical manipulation follows to "unbias" the estimates. Mathematically, the concept is based on

$$E_{P(x)}[\theta] = E_{g(x)}\left[\theta \cdot \frac{P(x)}{g(x)}\right], \tag{4.59}$$

where $g(x)$ is the distorted sampling function. Thus, the average of a variable $\theta(x)$ can be computed in two ways. The first corresponds to the left handside of (4.59) where we rely on sampling according to the natural probability $P(x)$. The second is by sampling $\theta(x) \cdot P(x)/g(x)$ according to a new distribution $g(x)$, the importance sampling function; the latter is nothing but weighting the sample points of $\theta(x)$ by a correction factor.

Therefore, in importance sampling, while the sample points are obtained from the importance sampling distribution, $g(x)$, the estimate is still based on the true distribution by applying weights to each observation. The weights stated in (4.60) are chosen according to (4.59). They are inversely proportional to the probability of outcomes under $g(x)$ compared to $P(x)$, and are often referred to as the "inverse likelihood ratio."

$$W(x) = \frac{P(x)}{g(x)} \tag{4.60}$$

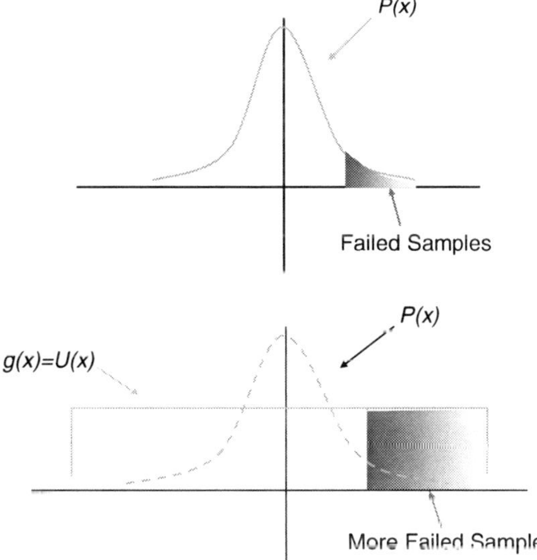

Fig. 4.15 Standard Monte Carlo generates many sample points around the mean. Example importance sampling where more sample points are generated around the tail

With the proper choice of $g(x)$, the method enables us to obtain reasonably accurate estimation results with relatively small number of sample points. To summarise, the importance of importance sampling is that it provides significant gains for the simulation of rare events; the latter being one of the most difficult statistical simulation problems.

In the following sections, we discuss three different methods to compute the importance sampling estimates: integration, ratio and regression methods [28]. This helps us understand the underlying principles of importance sampling and, hence, helps us make better judgement regarding the optimal choice of $g(x)$.

4.4.2.1 Integrated Importance Sampling

Given a random variable $\theta(x)$, with an unknown distribution, we want to estimate

$$\mu = E(\theta(X)) = \int \theta(x)P(x)\mathrm{d}x \tag{4.61}$$

$P(x)$ is the probability density function of x. Typical Monte Carlo analysis would estimate μ as

$$\hat{\mu}_{\mathrm{MC}} = \frac{1}{N}\sum_{i=1}^{N}\theta(x_i) \tag{4.62}$$

As we discussed earlier, in Sect. 4.3.3.1, the estimate is unbiased and asymptotically normally distributed.

By sampling from $g(x)$ and setting $Y(x) = \theta(x)\cdot P(x)/g(x)$, we can calculate

$$\hat{\mu}_{\mathrm{int}} = \frac{1}{N}\sum_{i=1}^{N}Y(x_i) \tag{4.63}$$

as an unbiased integrated importance sampling estimate. The ideal $g(x) = \theta(x)P(x)/\mu$ would require only one sample point to estimate the mean. However, it requires knowledge of the exact integrand, μ, itself.

Note that we can rewrite (4.63) as

$$\hat{\mu}_{\mathrm{int}} = \frac{1}{N}\sum_{i=1}^{N}W(x_i)\theta(x_i) \tag{4.64}$$

and

$$E_g[W(x_i)] = 1 \tag{4.65}$$

$W(x)$ is a random variable, and the weights $\{W(x_i)\}$, $i = 1,\ldots,N$, do not sum to one.

4.4.2.2   Ratio and Regression Estimates

The ratio estimate is obtained by normalizing the weights used in the integration estimate by some constant, such that the sum of the weights would add up to one. Equations (4.66) and (4.67) illustrate two equivalent forms of computing the ratio estimate.

$$\hat{\mu}_{\text{ratio}} = \frac{\sum_{i=1}^{N} W(x_i)\theta(x_i)}{\sum_{i=1}^{N} W(x_i)} \tag{4.66}$$

$$\hat{\mu}_{\text{ratio}} = \frac{\hat{\mu}_{\text{int}}}{\bar{W}} \tag{4.67}$$

While the ratio estimate tries to correct for the weights to have total mass of unity, the regression estimate corrects for the unity mass assumption and the expected value of W.

$$\hat{\mu}_{\text{reg}} = \sum_{i=1}^{N} \pi_i Y_i \tag{4.68}$$

π_i are "metaweights" [28]. Usually, the regression estimate is computed according to (4.69)

$$\hat{\mu}_{\text{reg}} = \bar{Y} - \hat{\beta}(\bar{W} - 1). \tag{4.69}$$

$\hat{\beta}$ is the value of the regression of Y on W at $W = 1$.

4.4.2.3   Variance and Confidence Intervals

According to [28], the estimates are asymptotically normally distributed. Hence, the $(1 - 2\alpha)$ confidence intervals can be derived according to $[\mu - c\sigma, \mu + c\sigma]$, where c satisfies $P(x > c) = (1 - \alpha)$, and σ is the standard deviation of the estimate. The standard deviation σ can be derived according to the following equations:

$$\hat{\sigma}_{\text{int}}^2 = \frac{1}{N(N-1)} \sum_{i=1}^{N} (Y_i - \hat{\mu}_{\text{int}})^2 \tag{4.70}$$

$$\hat{\sigma}_{\text{ratio}}^2 = \frac{1}{N(N-1)} \sum_{i=1}^{N} (Y_i - W_i\hat{\mu}_{\text{ratio}})^2 \tag{4.71}$$

$$\hat{\sigma}^2_{\text{reg}} = \frac{1}{N(N-2)} \sum_{i=1}^{N} (Y_i - \hat{\beta} W_i - \bar{Y} + \hat{\beta} \bar{W})^2 \tag{4.72}$$

If the weights are very small, some approaches correct for the confidence intervals by dividing the standard deviation of the estimate by the sample weight average.

4.4.2.4 Exponential Change of Measure

Traditionally, importance sampling deals with tail probabilities. It is possible to enable more probability mass in tails by scaling as follows:

$$g^*(x) = \frac{1}{\alpha} P\left(\frac{x}{\alpha}\right) \tag{4.73}$$

However, it has been shown that it is not very productive in complex high-dimensional problems. Instead, we often resort to translation for shifting the distribution to the failure region.

$$g^*(x) = g(x - T) \tag{4.74}$$

For normal distributions, this is equivalent to the exponential change of measure. Exponential change of measure is often referred to as exponential twisting or exponential tilting. An exponentially tilted function would have the form

$$g^*(x) = \frac{e^{\psi x} p(x)}{M(\psi)} \tag{4.75}$$

$$M(\psi) = E_p[e^{\psi x}] \tag{4.76}$$

The inverse likelihood ratio would be

$$W(x) = \frac{p(x)}{g(x)} = \frac{M(\psi)}{e^{\psi x}} \tag{4.77}$$

The exponential change of measure for a Gaussian $\sim N_g(\mu, \sigma^2)$ is $N_g(\mu - \psi, \sigma^2)$.

4.5 Importance Sampling for Memory Design

4.5.1 *The Importance Sampling Distribution*

A simple and logical choice of *g(x)* is the uniform distribution (Fig. 4.15). However, it was shown in [29] that its efficiency decreases as the dimensionality increases. Many techniques have been proposed to choose a good sampling distribution.

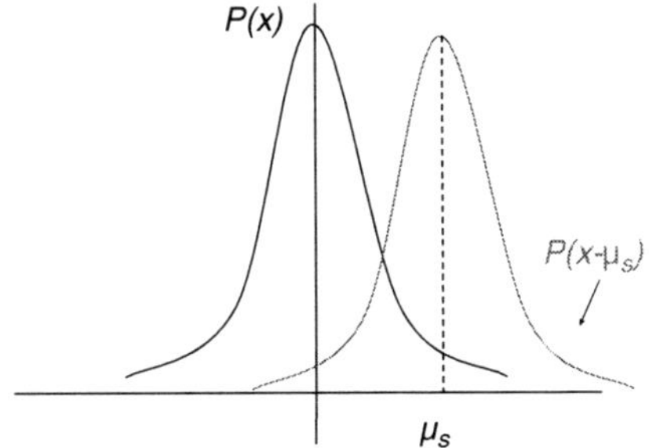

Fig. 4.16 Without loss of generality, $P(x)$ is a normal Gaussian in one-dimensional parameter space. $P(x - \mu_s)$ is $P(x)$ shifted to the new center μ_s

A common approach is to shift the natural distribution into the failure region [28]. Choosing $g(x) = P(x - \mu_s)$ as discussed in the previous section enables more sample points in the failure region (see Fig. 4.16).

The authors in [30] go one step further. Mixture ratioed importance sampling depends on the idea of generating random variables using mixtures of distributions [28]. Thus, $g(x)$ is chosen as follows:

$$g_\lambda(x) = \lambda_1 p(x) + \lambda_2 U(x) + (1 - \lambda_1 - \lambda_2)p(x - \mu_s), \qquad (4.78)$$

where $0 \leq \lambda_1 + \lambda_2 < 1$, and $0 \leq \lambda_1$ and $0 \leq \lambda_2$. $g_\lambda(x)$ enables focusing on the failure region without leaving any cold spots (i.e. any non-sampled regions in the event of outliers, etc.). Moreover, the method is generalisable to support multiple failure regions. The choice of λ_i is dependent on the location of μ_s; in general, if μ_s is far from the origin, λ_i is small. We will discuss the choice of μ_s in the next section.

The fact that $P(x)$ is part of the mixture distribution helps bound the weight function W (4.79) used in estimating the failure probability P_f (4.80); this in turn bounds $\sigma_{Pf:MixIS}$, the standard deviation of P_f, [28] (4.81). The number of sample points drawn from $g_\lambda(x)$ is N. $I(x)$ is the indicator function defined in (4.43).

$$W(x) = \frac{P(x)}{g_\lambda(x)} \qquad (4.79)$$

and

$$y(x) = W(x)I(x) \qquad (4.80)$$

$$\widehat{P}_f = \frac{\sum_{i=1}^{N} y(x_i)}{\sum_{i=1}^{N} W(x_i)} = \frac{\overline{y(x)}}{\overline{W(x)}} \qquad (4.81)$$

$$\widehat{\sigma}_{Pf:MixIS}^2 = \frac{1}{N} \frac{1}{N-1} \frac{1}{\overline{W(x)}} \sum_{i=1}^{N} \left(y(x_i) - \widehat{P}_f W(x_i)\right)^2 \qquad (4.82)$$

4.5.1.1   Shifted Mean Estimation

We propose the following heuristic for estimating the shift μ_s.

> **Pseudo Code: Estimating μ_s**
>
> 1. Uniformly sample the parameter space
>   a. Identify failing points
>   b. If (total number of failing sample points < 30)
>   Go to 1
> 2. Find the center of gravity (C.O.G.), or centroid of failures.
>   Set μ_s=C.O.G.

Note that a few failing sample points are in general sufficient for estimating the centroid (COG) because we are estimating a mean. Furthermore, another round of sampling follows this step; the estimates obtained by importance sampling are not highly sensitive to the exact position of the COG, or μ_s. This is true as long as the uniform sample points in step "1" are good representatives of their corresponding uniform population.

To obtain an idea about the number of uniform sample points that would typically be needed, let us carry out the following experiment. In this experiment, we assume that in general we are interested in obtaining fail probabilities whose equivalent z_0, as studied in Sect. 4.3.4 (Table 4.2), would be in the neighborhood of 5.5. The reason for picking this number is that designers often target fails of complementary cumulative probability of 1-per-million; $z_0 = 5.5$ in a standard normal distribution corresponds approximately to 1-per-50 million. Figure 4.17 illustrates a two-dimensional space example with a linear fail boundary. The illustrated example shows the case when all variables (dimensions) contribute to the fails. We also study the case when only one variable determines the fail criteria. We also assume that the search for the uniform sample points will fall in a cube whose bounds are $[-m\sigma, +m\sigma]$. For this example, we pick $m = 6$.

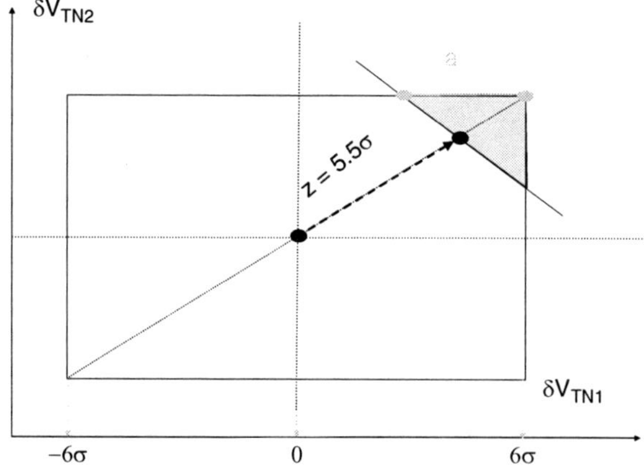

Fig. 4.17 Possible fail boundary in two-dimensional space

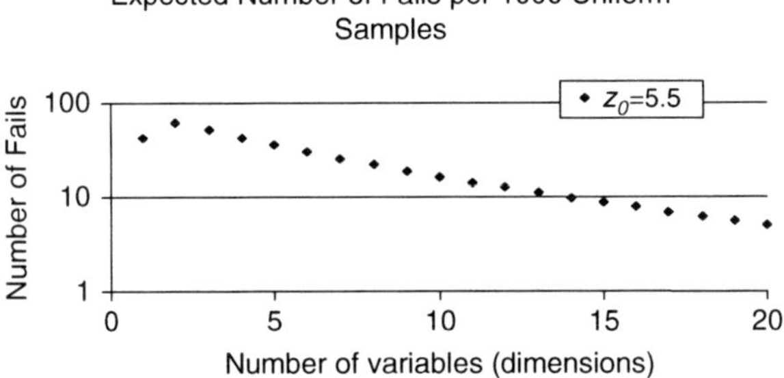

Fig. 4.18 Number of failing sample points for hypothetical fail region (Fig. 4.16)

The ratio of the volume of fail region to the volume of the cube will be proportional to the number of failing sample points if we uniformly sample the space. Figure 4.18 illustrates the expected number of failing sample points per 1,000 uniform sample points as a function of the dimensionality of the problem. We notice that for medium-sized problems like SRAM cells, we can detect the COG with few thousand simulations.

To guarantee that the uniform sample points in step "1" span the parametric (variability) space properly, we can employ Sobol' sequences (see Sect. 4.4.1.3). The sample points from such sequences are "maximally avoiding" of each other [16] and hence are expected to provide good coverage of the sampled cube.

Exploiting knowledge of circuit behavior: typically, the uniform sampling range would span $[-m\sigma, +m\sigma]$ range to provide good coverage of the fails. An example of $m = 6$ would still allow one to catch a 5σ fail even if the fail boundary is governed by one variable. In certain applications, where designers have apriori knowledge of the design, it is possible to restrict the COG search range based on known characteristics of the circuit behavior for a given dimension. For example, let us say that we know that design will fail only if a transistor is weak (i.e. if its corresponding threshold voltage Vt increases, i.e. $\delta Vt > 0$). In that case, there is no need to span the full strong range (negative δVt). One can pick the range to be $[-2\sigma, 6\sigma]$ versus $[-6\sigma, 6\sigma]$. This can enhance the search efficiency of the COG; maintaining a negative margin enables good guard against human error.

4.5.1.2 Theoretical Application

Let us repeat the experiments in Table 4.2, this time for the mixture importance sampling method (MixIS) described in Sect. 4.5.1. We rely on (4.83), the integral form of (4.82), to calculate the number of MixIS sample points N_{IS} needed to estimate $P_f = \text{Prob}(x > z_0)$ with a 95% confidence interval and error criterion $\gamma = 0.1$ (see (4.45)). Again x follows a standard normal distribution. Results show that the MixIS method provides significant speedup for low failure probabilities; the method matches regular Monte Carlo when the latter converges for high probability estimates (see Table 4.5). Figure 4.19 illustrates the speed improvement due to importance sampling. The gain increases as we estimate probabilities of events that occur further down the tails.

$$\sigma^2_{Pf:\text{MixIS}} = \frac{1}{N_{\text{IS}}} \int \left(y(x) - \hat{P}_f w(x)\right)^2 g(x)\,dx \tag{4.83}$$

Let us now test the efficiency of the MixIS method in higher dimensions. For this, we rely on experimentation (Fig. 4.20). We create a simple function, $f(\mathbf{x})$, for

Table 4.5 The number of Monte Carlo sample points (second column) and MixIS sample points (third column) needed to estimate $P_f = \text{Prob}$ $(x > z_0)$ with 95% confidence interval $= [P_f - 0.1P_f, P_f + 0.1P_f]$

z_0	N_{MC} # Monte Carlo simulations	N_{IS} #MixIS simulations	Speedup $= N_{\text{MC}}/N_{\text{IS}}$
0	4.0e2	4.0e2	1e0
1	2.1e3	5.8e2	4e0
2	1.7e4	9.3e2	2e1
3	2.9e5	1.4e3	2e2
4	1.3e7	1.8e3	7e3
5	1.4e9	2.3e3	6e5

Speedup of MixIS compared to regular Monte Carlo (fourth column)

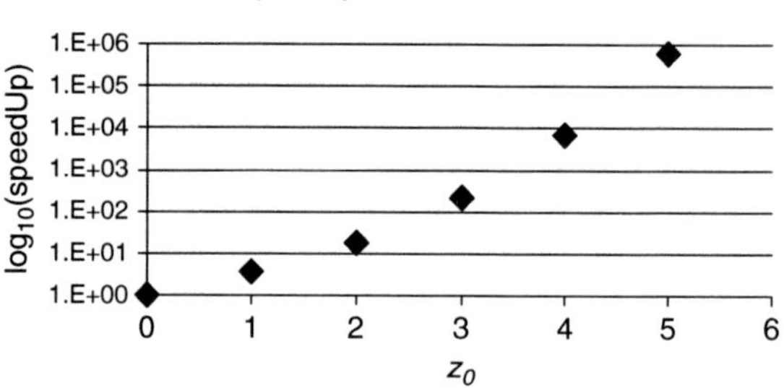

Fig. 4.19 Speedup of proposed method compared to Monte Carlo

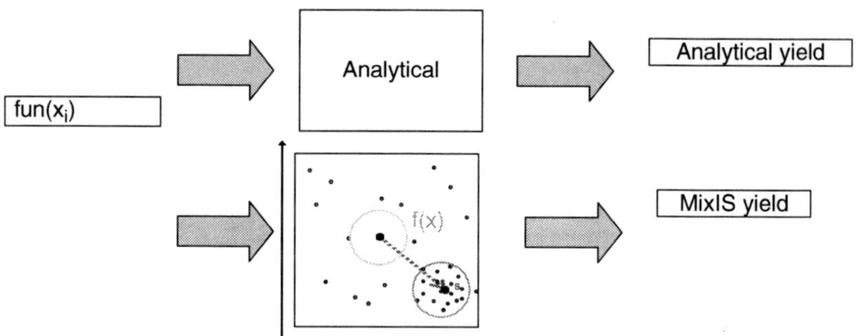

Fig. 4.20 Illustration of framework for comparison against a function whose analytical yield is easy to compute

which the yield could be computed analytically. We then compare the MixIS estimate to the computed analytical yield value. We represent the analytical and estimated yields in terms of equivalent $z_{0-\mathrm{eqv}}$ values, which are related to the probability of fail P_f according to (4.84).

$$
\begin{aligned}
P_\mathrm{f} &= P(f(\mathbf{x}) > f^0) = P(z > z_{0-\mathrm{eqv}}) \\
z_{0-\mathrm{eqv}} &= \varphi^{-1}(1 - P_\mathrm{f}) = \varphi^{-1}(1 - \mathrm{Prob}(f(\mathbf{x}) > f_0));
\end{aligned}
\tag{4.84}
$$

$z \sim N_\mathrm{g}(0,1)$, φ is the standard normal cumulative distribution function, and f^0 is the pass/fail critical value. We then use MixIS to estimate $z_{0-\mathrm{eqv}}$ for different f^0 values. Our objective is to determine the number of MixIS sample points needed to obtain a good estimate. Table 4.6 presents the results for a six-dimensional space experiment; note that we are interested in six-dimensional space because SRAM designs typically deal with six random variables. The confidence intervals are based on fifty replications, and the number of MixIS sample points per replication was fixed to 1,500. The method, thus, maintains its efficiency compared to regular Monte Carlo techniques, which require a much larger number of sample points as indicated in Table 4.5. Furthermore, MixIS is found to be more efficient than other candidate importance sampling functions; for example, using $U(x)$ alone loses efficiency as the dimensionality increases. It is worth noting that a six-dimensional $U(x)$ requires at least 10–15× more sample points to obtain a good estimate.

4.5.2 SRAM Application

The threshold voltage fluctuations of the SRAM cell transistors (Fig. 4.21) are impacted by the random dopant fluctuations and may be considered as six independent Gaussian random variables, δVt_i, $i = 1$–6 [30]. Process variations between the neighbouring transistors can degrade the cell performance and stability.

Table 4.6 Comparing the MixIS method against analytical solutions for a six-dimensional parameter space

Analytical $z_{0-\mathrm{eqv}}$	95% Confidence interval lower bound	MixIS $z_{0-\mathrm{eqv}}$	95% Confidence interval upper bound
3.0	2.94	2.99	3.09
3.5	3.44	3.49	3.57
4.0	3.96	4.01	4.06
4.5	4.45	4.48	4.55
5.0	4.96	4.99	5.07
5.5	5.45	5.52	5.55
6.0	5.96	5.99	6.07

The second (fourth) column presents the confidence interval lower (upper) bound values for 50 replications. The third column presents the estimated mean value. Each replication involved 1,500 simulations

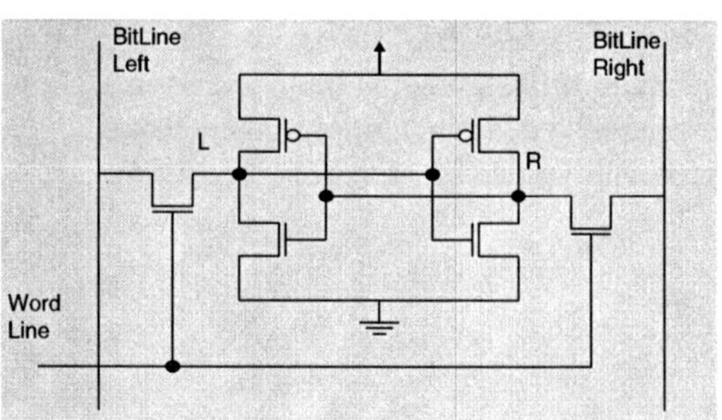

Fig. 4.21 Schematic of the 6T SRAM cell under study

For state-of-the-art memory chips with million or more cells, this can impact the yield of the design significantly. Moreover, the yield estimate is very sensitive to estimating the probability of fail accurately. For example, the yield estimate can drop from 90 to 40% when $z_{0-\mathrm{eqv}}$ drops from 5.3 to 4.8. Estimating such extreme tail probabilities with good accuracy is not practical using traditional Monte Carlo. In the following examples, we will see how the MixIS method lends itself as an efficient statistical methodology to estimate such high yields (low failure probabilities). First, we will study the SRAM cell metrics.

4.5.2.1 Dynamic Stability and Writability for PD/SOI Designs

While it has been the norm to rely on the static noise margin [31] for statistical analysis, this technique is no longer sufficient. This is particularly true for high-performance applications, and transient analysis is needed to evaluate the cell functionality and performance margins. Namely, for PD/SOI devices [32], the operating conditions (i.e., V_{ds} and V_{gs} values) of a device modify the threshold voltage due to the floating body effect. Moreover, the read/write operation of a cell also depends on the capacitances at the cell nodes, which can only be captured through transient analysis [33].

Hence, we study the SRAM cell functionalities in terms of its dynamic behavior: (1) cell dynamic stability, or the ability to maintain cell status during read "0"; e.g., V_{max} noise in Fig. 4.22 does not exceed $Vdd/2$ and (2) cell writability, or the ability of writing a "0" to the cell. One can also define the cell read delay to be the time, t_{read}, taken by the cell to discharge the bitline below a given threshold.

4.5.2.2 Yield Analysis

We used the MixIS method to estimate the yield of a six-transistor SRAM cell built in sub-100 nm technology in the presence of variability. We are interested in measuring the cell's dynamic stability in terms of read upsets and writability. For practical analysis purposes, the circuit under study consists of the following

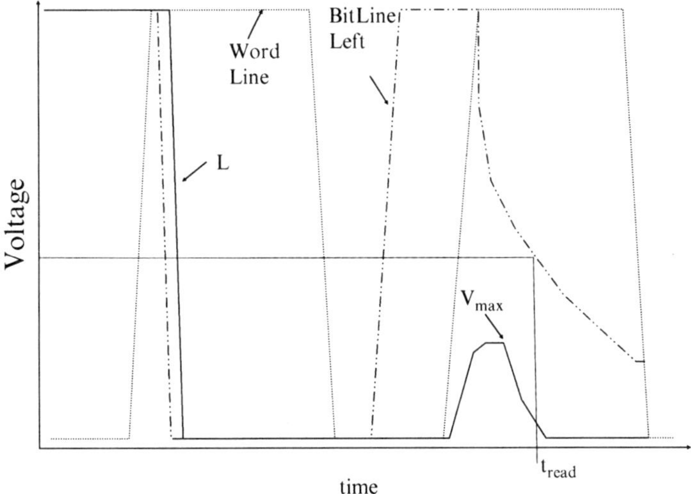

Fig. 4.22 SRAM cell write "0" followed by a read "0." Refer to Fig. 4.21

components: the SRAM cell, bitline loading and peripheral circuitry. We rely on the following indicator functions to estimate the yield.

$$I(\delta Vt_1, ..., \delta Vt_6) = \begin{cases} 0, \text{stable cell} \\ 1, \text{read upset(write\_fail)} \end{cases} \tag{4.85}$$

The overall yield is computed from the individual (metric) yields. For purposes of our experiments, we report the cell yield as an equivalent z_{0-eqv} as described in (4.84). The device threshold voltages were normally distributed with zero mean, and their standard deviation extrapolated from hardware.

Figure 4.23 compares the MixIS method to regular Monte Carlo when estimating the cell yield. Both methods converge to the same estimated value. MixIS converges quickly, with few thousand simulations, whereas Monte Carlo is very slow when dealing with rare failure events (very high yields). Table 4.7 compares estimated z_{0-eqv} values obtained via MixIS to those obtained by regular Monte Carlo, whenever a converging Monte Carlo is realizable. MixIS estimates are in excellent agreement with Monte Carlo results. A converging MixIS estimate was achieved with $\sim 2,000$–$3,000$ sample points, regardless of the z_{0-eqv} value. Whereas the number of Monte Carlo sample points increased exponentially with z_{0-eqv}, for $z_{0-eqv} > 4$, it was no longer practical to rely on regular Monte Carlo. Most importantly, MixIS is a computationally efficient alternative to regular Monte Carlo, and the runtime is independent of the yield estimate. This makes it a suitable methodology for accurately estimating rare fail probabilities of SRAM. Finally, in Table 4.8 we compare MixIS method results to hardware collected statistics. The results and the trends are in very good agreement. Some discrepancy is seen largely due to the mismatch between the device models used for simulation and the true hardware behavior.

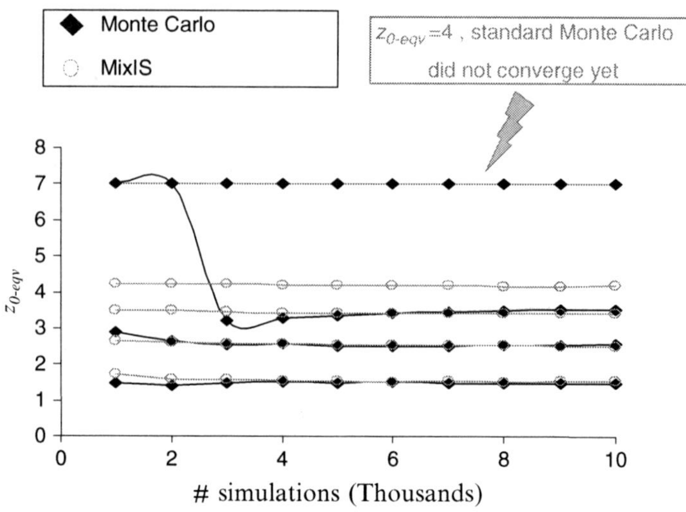

Fig. 4.23 $z_{0-\mathrm{eqv}}$ estimated using both MixIS (circles) and Monte Carlo (diamonds) techniques. For all $z_{0-\mathrm{eqv}}$ values MixIS reached a converging estimate and a converging confidence interval within few thousand simulations. The Monte Carlo method fell behind for higher $z_{0-\mathrm{eqv}}$, i.e. cells with rare failure events. Around 100,000 simulations were needed for regular Monte Carlo to provide a converging estimate for $z_{0-\mathrm{eqv}} \approx 4$. Many more sample points are necessary for Monte Carlo method to satisfy the converging confidence interval criteria

Table 4.7 Estimated $z_{0-\mathrm{eqv}}$ values

Monte Carlo	MixIS
1.49	1.53
2.56	2.51
3.49	3.42
4.15	4.22
–	4.96
–	5.68
–	6.06

MixIS provided a converging estimate within 2,000–3,000 simulations. Monte Carlo requirements exceeded 100,000 simulations for $z_{0-\mathrm{eqv}} > 4$. Monte Carlo results are based on 100,000 sample points

Table 4.8 Log of the number of fails based on hardware measurements and those estimated by the MixIS method

Supply voltage	Vdd_1	Vdd_2	Vdd_3	Vdd_4	Vdd_5	Vdd_6
Hardware	−1 to 0	0–1	1–2	3–4	4–5	5–6
MixIS	0–1	0–1	2–3	3–4	4–5	4–5

4.5.2.3 Effects of Supply Fluctuation

Typically, the designers would use the statistical sampling methodology to predict yield trends as a function of the supply voltage (Fig. 4.24). In this section, we

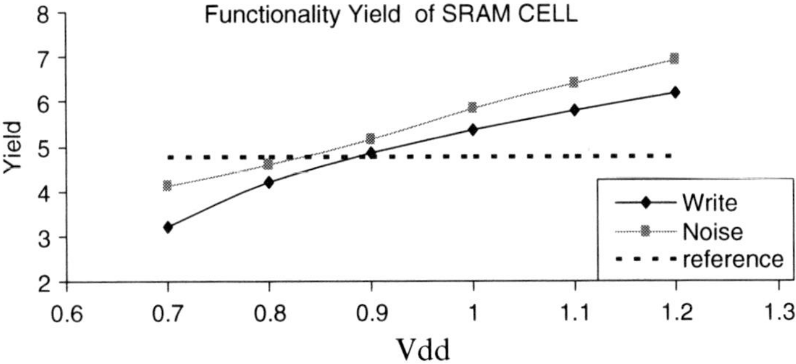

Fig. 4.24 SRAM cell yield for writability and stability

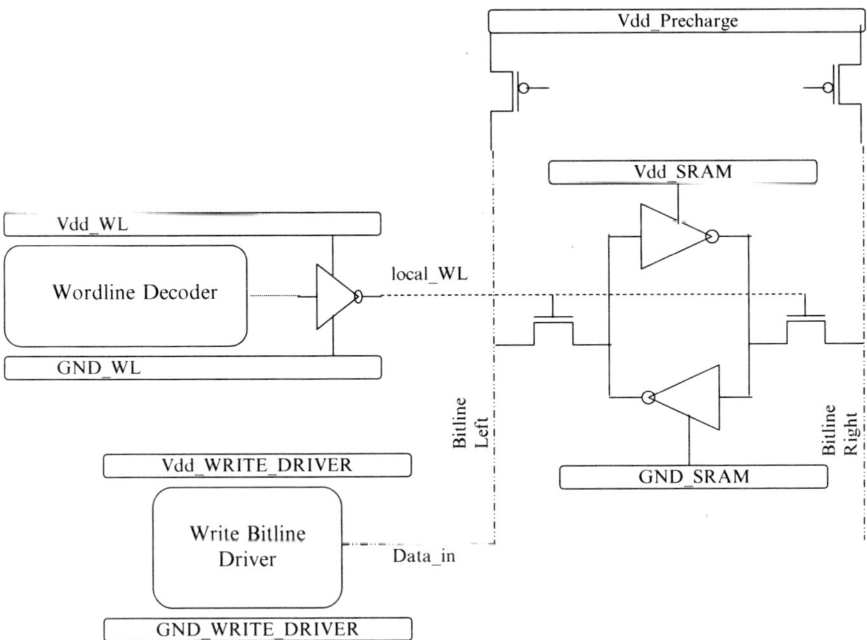

Fig. 4.25 Example cross section of an SRAM cell

extend this to analyse cell supply sensitivities. Figure 4.25 presents an example of
an SRAM cell cross-section of interest. Note that the cell and periphery have
separate supplies. It is possible with supply fluctuations that the balance between
cell and peripheral circuitry gets upset leading to increased delay or functional fails.
Because we are dealing with small fluctuations (5–10%) [34], we can rely on the
sensitivity of a given metric to the supply fluctuations to represent the metric
dependence on the supply variation. We generalise the definition in [35] for sensitivity according to (4.86).

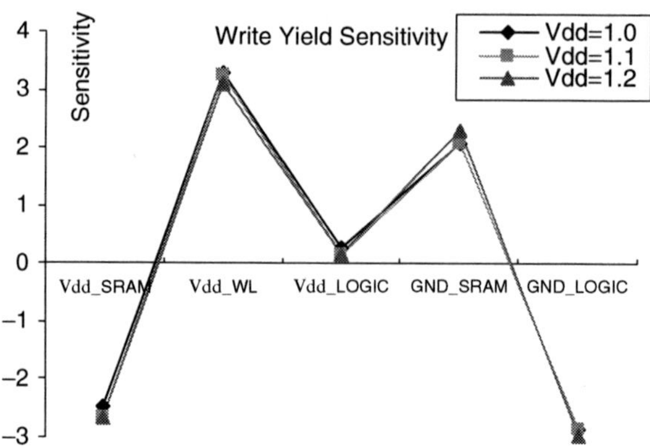

Fig. 4.26 Sensitivity of cell writability yield, $S_{V_{dd}(GND)\_i}^{\text{yield}}$, to fluctuations in the different supplies

$$S_{Vdd}^{\text{metric}} = \lim_{\Delta Vdd \to 0} \frac{\frac{\Delta\text{metric}}{\text{metric}}}{\frac{\Delta Vdd}{Vdd}} = \frac{Vdd}{\text{metric}} \cdot \frac{d\text{metric}}{dVdd} \tag{4.86}$$

To better quantify the combined effect of supply fluctuation and threshold voltage fluctuation, we study the sensitivity of cell functionality yield to these fluctuations. We then model the yield sensitivity to predict ranges of tolerable fluctuations as functions of desired $V_{\min}$ (minimum cell operating voltage for given tolerance level).

$$\frac{\Delta\text{yield}}{\text{yield}} = \sum_{i=1:7} a_i \Delta x_i \tag{4.87}$$

Δx_i correspond to the relative fluctuations in the ith supply, $\frac{\Delta \mathbf{Vdd_i}}{\mathbf{Vdd_i}}$; a_i are the fluctuation coefficients derived from the individual linear sensitivities in (4.86), $S_{Vdd_i}^{\text{Yield}}$. Hence, (4.87) is a linear regression based on one-factor-at-a-time sensitivity analysis. Figure 4.26 presents the estimated cell writability yield sensitivity to different sources of supply fluctuation.

4.6 Conclusions

In this chapter, we reviewed importance sampling-based yield estimation methods for memory designs. Our goal is to introduce the concept of "smart" sample generation methods as an alternative to traditional Monte Carlo through learning and control of the designs. We hope that the background information was insightful and that we left the reader with ambitions to explore more of the beautiful world of probability, statistics and the theory of chance.

References

1. Willcox W (1938) The founder of statistics. Rev Int Stat Inst 5(4):321–328
2. Apostol TM (1969) Calculus, Volume II, 2nd edn. Wiley, New York
3. Two Sciences, http://www.ortho.lsuhsc.edu/Faculty/Marino/Point1/Representative.html
4. Cochran WG (1977) Sampling techniques, 3rd edn. Wiley, New York
5. Park S, Miller K (1988) Random number generators: Good ones are hard to Find. Commun ACM 31(10):1192–1201
6. Knuth DE (1981) The art of computer programming, 2nd edn. Addison-Wesley, Reading, MA
7. Matsumoto M, Nishimura T (1998) Mersenne twister: A 623-dimensionally equidistributed uniform pseudo-random number generator. ACM Trans Model Comput Simul 8(1):3–30
8. Le Cam L (1986) The central limit theorem around 1935. Stat Sci 1(1):78–91
9. Box GE, Muller ME (1958) A note on the generation of random normal deviates. Ann Math Stat 29(2):610–611
10. Devroye L (1986) Non-uniform random variate generation. Springer, New York
11. George M, Wan TW (2000) The Ziggurat method for generating random variables. J Stat Software 5(8):1–7
12. Walpole R, Myers R, Myers S (1993) Probability and statistics for engineers and scientists, 2nd edn. Macmillan, New York
13. Haugh, M. The Monte Carlo framework, examples from finance and generating correlated random variables. http://www.columbia.edu/~mh2078/MCS04/MCS_framework_FEegs.pdf
14. Hammersley JM, Handscomb DC (1964) Monte Carlo methods, Fletcher and Son Limited, Norwich
15. Metropolis N (1987) The beginning of the Monte Carlo method. Los Alamos Science, Special Issue dedicated to Stanislaw Ulam, pp 125–130. http://library.lanl.gov/la-pubs/00326866.pdf
16. Press WH, Flannery BP, Teukolsky SA, Vetterling WT (1988) Numerical recipes in C. Cambridge University Press, Cambridge, UK
17. Judd KL (1998) Numerical methods in economics. MIT Press, Cambridge, MA
18. McKay MD, Beckman RJ, Conover WJ (2000) A comparison of three methods for selecting values of input variables in the analysis of output from a computer code. Technometrics 42 (1):55–61, (1979) Am Stat Assoc 21(2):39–245
19. Iman RL, Helton JC, Campbell JE (1981) An approach to sensitivity analysis of computer models, Part 1. Introduction, input variable selection and preliminary variable assessment. J Quality Technol 13(3):174–183
20. Box GE, Hunter WG, Hunter JS (1978) Statistics for experimenters: An introduction to design, data analysis, and model building. Wiley, New York
21. Rousond M, Hoyland A (2003) System reliability theory: Models, statistical methods, and applications. 2nd edn. Wiley-Interscience
22. Simon JL (1995) Resampling: The new statistics. resampling stats. http://www.resample.com/content/text/index.shtml
23. Henderson S, Nelson, H (eds) (2006) Handbooks in operations research and management science, vol 13. Sampling, North-Holland, Amsterdam
24. Glasserman P (2003) Monte Carlo methods in financial engineering. Springer, New York
25. Ross SM (2006) Simulation, 4th edn. Academic Press, New York
26. Niederreiter H (1992) Random number generation and quasi-Monte Carlo methods. Society for Industrial and Applied Mathematics, Philadelphia, PA
27. Thiemard E (2001) An algorithm to compute bounds for the star discrepancy. J Complex 17 (4):850–880
28. Hesterberg TC (1988) Advances in importance sampling. Ph.D. Dissertation, Statistics Department, Stanford University
29. Hocevar DE, Lightner MR, Trick TN (1983) A study of variance reduction techniques for estimating circuit yields. IEEE Trans Comput Aided Des Integrated Circuits Syst 2(3): 180–192

30. Kanj R, Joshi R, Nassif S (2006) Mixture importance sampling and its application to the analysis of sram designs in the presence of rare failure events. Des Autom Conf 69–72
31. Seevinck E, List FJ, Lohstroh J (1987) Static noise margin analysis of MOS SRAM cells. IEEE J Solid-State Circuits SC-22(5):748–754
32. Narasimha S et al (2006) High performance 45-nm SOI technology with enhanced strain, porous low-k BEOL, and immersion lithography. IEEE Int Electron Device Meet 1–4
33. Joshi RV, Mukhopadhyay S, Plass DW, Chan YH, Ching-Te Chuang YT (2009) Design of sub-90 nm low-power and variation tolerant PD/SOI SRAM cell based on dynamic stability metrics. J Solid-State Circuits 44(3):965–976
34. Kanj R, Joshi R, Nassif S (2007) SRAM yield sensitivity to supply voltage fluctuations and its implications on V_{min}. Integr Circuit Des Technol Conf 1–4
35. Alioto M, Palumbo G (2006) Impact of supply voltage variations on full-adder delay: analysis and comparison. IEEE Trans Very Large Scale Integration 14(12):1322–1335

Chapter 5
Direct SRAM Operation Margin Computation with Random Skews of Device Characteristics

Robert C. Wong

5.1 Introduction

SRAM has been generally characterized with some SNM [12] from the voltage–voltage (VV) plots or the Icrit from the current–voltage (IV) plots. They do indicate the robustness of the SRAM operations but would not provide sufficient information for SRAM designers, as to the possible SRAM yield and the redundancy requirements. One way to estimate SRAM yield is based on the expected fail count with the Poisson distribution

$$\text{YIELD} = \sum_{n=0}^{k} \frac{\lambda^n \exp(-\lambda)}{n!} \tag{5.1}$$

where k is the maximum number of fails in the chip that can be repaired

λ is the expected fail count of the chip

Traditional method to estimate the expected fail count is to run Monte Carlo simulations. The fail count then defines the λ of a particular SRAM design, which in turn defines the amount of redundant bits needed to meet the yield target. Running Monte Carlo simulation with random number generators is in a way giving up on the analytical reasoning because the overall situation seems too complex to reason through. The basic difficulty of Monte Carlo simulation is the computer run time. To estimate one fail per billion, more than one billion runs are needed which would overwhelm the common computing resource. Fail count <1 ppb (parts per billion) would be needed to yield 96% of 64 M full good SRAM. Another difficulty is to keep track of the various device parameters which affect the SRAM operations, especially in the early phase of a technology node when the parameters are still

R.C. Wong
IBM Microelectronics Division, Hopewell Junction, NY, USA
e-mail: wongr@us.ibm.com

A. Singhee and R.A. Rutenbar (eds.), *Extreme Statistics in Nanoscale Memory Design*,
Integrated Circuits and Systems, DOI 10.1007/978-1-4419-6606-3_5,

drifting. Thus, even if the computing resource is available and timely enough for the Monte Carlo runs, there is still great uncertainty on the distributions of the various parameters. It will take a long time for the distributions to settle. By then, actual hardware data is usually available for design refinements. The prediction of SRAM fail count from Monte Carlo runs would be less urgent, and so model-hardware correlation is generally neglected. Direct margin computation is to estimate SRAM operation margins in a more analytical manner with only one to two parameters to be tracked and correlated along the evolution of a technology node. Timely and accurate margin computation is necessary for design and process adjustments to meet the scaling qualification targets.

5.2   General Metrics for SRAM Operation Margins

Some simpler SRAM ratios have also been used to indicate the SRAM functionality besides the metrics of SNM and Icrit. The β ratio of the SRAM pulldown Ion and passgate Ion is well known to correlate to stability. The γ ratio of passgate Ion and pullup Ion is also a general indicator for SRAM writability. The α ratio of pullup Ion and pulldown Ion correlates to the SRAM inverter trip point that indicates the threshold of cell flipping to the opposite polarity. Thus, higher α ratio implies better stability and writability. These ratios are less precise than the common metrics of SNM and Icrit. For example, higher β ratio may not lead to better stability if the higher β ratio is from faster pulldown, which also drops the α ratio. However, they do provide simple indicators of SRAM operations that can be easily monitored in the course of SRAM development. More precise metrics can be formed by the combination of the SRAM ratios. For example, a new stability metric can be formed as follows

$$\text{Vcrit} = (\text{Vtrip} - \text{Vzero}) = a\alpha - \frac{1}{b\beta} \tag{5.2}$$

where Vzero is the "0" node level as the passgate is turned on by the word line

Vtrip is the trip point of the inverter formed by the pullup and pulldown
b is some fitting constant for Vzero which is proportional to $1/\beta$
a is some fitting constant for Vtrip which is proportional to α

A similar metric for writability can also be formed as follows

$$\text{Vcritw} = (\text{Vtrip} - \text{Vzerow}) = a\alpha - \frac{1}{g\gamma} \tag{5.3}$$

where Vzerow is the voltage level between pullup and passgate when they are configured as a potential divider as in the write operation.

More precise Vtrip and Vzero can be acquired from the SRAM IV plot as shown in Fig. 5.2, where Vzero is 0.13 V and Vtrip is 0.42 V. Vcrit, the stability metric against access disturb is then 0.29 V. Those fitting constants would be closer to one if the SRAM ratios of Ion at Vds (drain-source voltage) = Vdd are refined to fit the actual operation condition. For example, β ratio can be refined to Ipdβ/Ipgβ, where Ipdβ is pulldown Ion at Vds = Vzero around 0.13 V, and Ipgβ is passgate Ion at Vds = Vdd − Vzero.

Icrit is defined by the IV plot or Ncurve of an SRAM cell. As SNM is acquired from voltage sweep on one cell node and voltage measurement on the other node, Icrit is acquired from voltage sweep on one cell node and current measurement on the same node. Some example IV plots are illustrated in Figs. 5.1–5.3. The device symbols are to indicate SOI (Silicon On Insulator) devices in the plot. Ncurve of a standby cell with word line down and bit line up is shown in Fig. 5.1. As the left cell node is swept up to Vdd, the Source Monitor Unit (SMU) has to supply current to raise the "0" node until it trips over around 0.38 V. After that, the SMU has to drain current from the cell node which is being pulled up to Vdd by the left side pullup device. If the SMU is disconnected before the Vtrip of 0.38 V, the cell node will settle back to GND. If the SMU is disconnected after the Vtrip, the cell node will settle up to Vdd. Thus, Icritu is maximum leakage current to the "0" node that the SRAM can tolerate before disturb. Icritd is the maximum tolerable leakage current from the "1" node.

The more urgent challenge in SRAM scaling is the stability against access disturb when the word line is up. Access disturb margin is diminishing with the smaller devices of bigger variability. When the word line is activated to access the cell data, the whole row of memory cells are to be disturbed. The cell "0" node is perturbed to Vzero $\sim$ 0.13 V by the potential divider of pulldown and passgate as

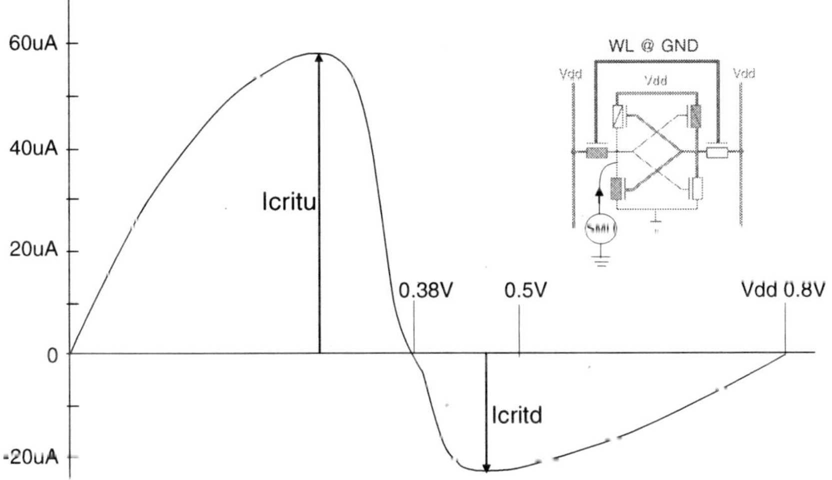

Fig. 5.1 Standby SRAM IV plot from current measurement during voltage sweep

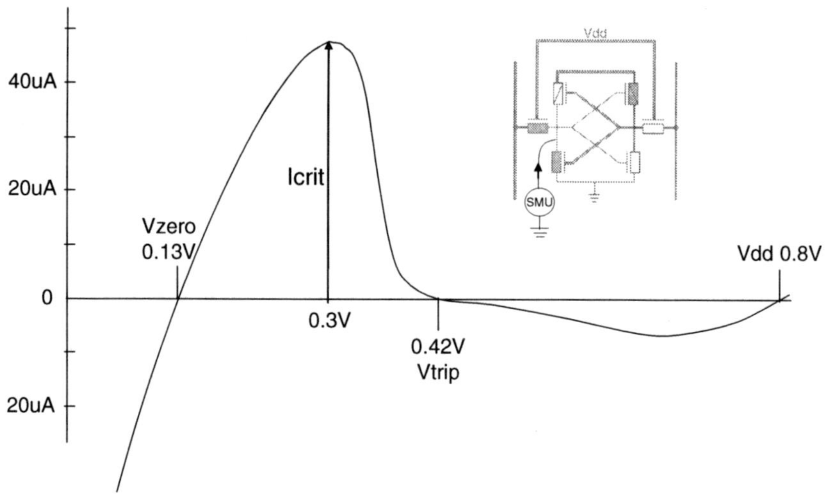

Fig. 5.2 IV plot with active word line for Icrit to compute stability fail count

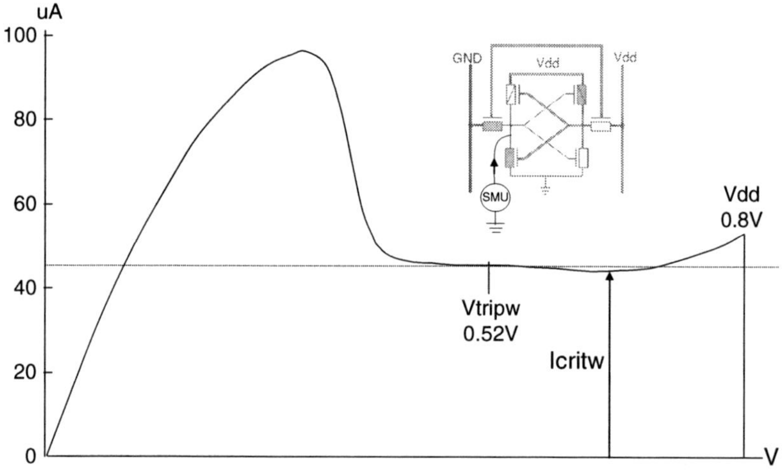

Fig. 5.3 IV plot with active word line for Icritw to compute write fail count

in the example of Fig. 5.2. As long as Vzero is below the trip point Vtrip, the cell is stable. Thus, stability metric can be defined by Icrit, the maximum tolerable leakage current to the "0" node. Another metric is Vcrit = Vtrip − Vzero, for the maximum tolerable voltage bump. A more precise metric would be Pcrit, the maximum tolerable disturb power computed from the area of the IV plot between Vzero and Vtrip.

It can be seen from Fig. 5.2 or (5.2) that higher SRAM β ratio does not necessarily lead to better stability. Higher β ratio does lower Vzero, but it may also lower Vtrip if the α ratio is also lowered by the faster pulldown. Actually, this suggests one opportunity to tune the cell for better stability with slower pulldown to reduce standby power dissipation and faster passgate to increase access current. Some more discussion about stability versus β ratio is in Sect. 5.7.

In low power SRAM cells with slower devices, writability may become harder to achieve than stability. One writability metric from the IV plot under the write conditions is illustrated in Fig. 5.3.

As the cell "1" node is swept down from Vdd, the SMU current must not drop below zero. The minimum current Icritw is then the writability metric. If Icritw is above zero, the "1" node will settle to "0" when the SMU is disconnected. Thus, the cell is writable to "0". If Icritw is below zero, the "1" node would be "stuck" at the first crossing with the zero current axis when it is swept down from Vdd. If the perturbation of the SMU is removed in the neighborhood of this crossing, the "1" node will settle to that crossing voltage while it is supposed to be pulled down all the way to GND by the left bit line. Thus, the cell is not writable.

SRAM fail count can then be estimated by the statistics of the particular metric. For example, if Icrit drops below zero at 6 σ process skew, the expected stability fail count is 2.07 ppB, which corresponds to about 87% stability yield for all good 64 M SRAM from (5.1). The fail count is from both sides of the normal density function beyond 6 σ. Here, 1 M stands for $(1024)^2$ and ppB is parts per binary billion of $(1024)^3$.

In the discussion below, stability fail count is to be represented by ADM for Access Disturb Margin in units of standard deviation of the normal density function. In a similar manner, write fail count computed from Icritw is represented by the WRM for WRite Margin. Standby data retention fail count computed from Icritu and Icritd is represented by DRM for Data Retention Margin. Some other margin relevant to SRAM operations is RTM for Read Time Margin. SRAM soft errors or single event upsets would be mainly related to DRM which can be computed with Vt skews plus some random leakage variations representing the effects of external upsets. Margins in σ are easier to see on linear plots than fail counts in ppm (parts per million) or ppb (parts per billion) on logarithmic plots. A sample look up table for SRAM margin targets to meet yield requirements is appended at the end of this chapter. SRAM Vmin or Vmax spec is based on the margin targets needed to provide reasonable yield for the particular technology node at the particular development cycle. For example, the margin target for the early exploration of 45 nm node SRAM cells may be set to 5.2 σ as shown in Sect. 5.6.

5.3 Idealization of Statistical Chaos in Single Variable

In Monte Carlo simulations, many device or process parameter values are randomly assigned within some statistical distributions. SRAM behavior are sensitive to many parameters such as Vt, Tinv (electrical gate insulator thickness), Toxgl (physical gate insulator thickness), carrier mobility, diffusion resistance, various components of gate capacitance and gate leakage, diffusion junction capacitance and leakage, lithographic misalignment, polysilicon line edge roughness, and actual device dimensions. They add up to more than a dozen parameter distributions that have to be monitored and tightened in the course of technology development. Monitoring all those parameters is very time consuming and is usually lagging behind the schedule of technology development. In the scaled down nodes, most parameter distributions cannot be properly acquired in time and are just extrapolated from the previous node, with the tolerance somewhat scaled for the new node. Unfortunately, the extrapolation from the previous node can be quite far away from reality because the many new process elements such as stress liners, self-aligned contacts, high K metal gate versions, embedded SiGe and SiC which are needed to keep scaling alive. Actually, extrapolation within the same node can be very far off when process tweaks are being applied from one wave of hardware to the other. Even when accurate parameter distributions become available, the millions of cases in Monte Carlo simulations can be unbearable burden on the available computing resource. Direct SRAM margin computation is an attempt to estimate operation margins from the distribution of a single parameter which can be diligently monitored and correlated with the margin computation which is to guide SRAM development to meet application spec.

The one single parameter that is most relevant to SRAM operations is the device threshold voltage Vt. Vt is also a sensitive function of most other device and process parameters relevant to SRAM functions. Variations in most parameters would directly reflect on Vt variations. Hardware correlation and model verification can thus be much simpler and more affordable. An alternative lump parameter for margin computation is device Ion. One problem is with Ion mismatch measurement in the device linear region, where source and drain voltage are very close. Current is very low and accurate monitoring become difficult, especially for low power devices in routine manufacturing measurements. This single random variable computation can be extended to include one or two extra random variables and remain manageable. For example, defect leakage can be included as the extra independent random variable to evaluate DRM.

5.4 Formulation of the Direct Computation for SRAM Margins

The premise for the direct computation is that one σ metric can be idealized in the root mean squares of the weighted standard deviations of device Vt variation. Device Vt variation is selected as the independent variable because it reflects closely the lump variation of most other parameters.

$$1\ \sigma\mathrm{M} = \sqrt{\sum_{i=1}^{6} G_i^2 \sigma_i^2} \qquad (5.4)$$

1 σM is the standard deviation of the metric such as Icrit or SNM.
$G_i = |\partial M/\partial x_i|$ is the absolute gradient value of M with respect to Vt skew
σ_i is the standard deviation of the device Vt

The Vt skews in the six transistors of the 6 T SRAM cell are also assumed to be in the "wrong" direction that degrades the metric. One first order estimate of the operation margin is then

$$\text{Operation Margin} = M0/1\sigma M \tag{5.5}$$

M0 is the nominal metric at nominal Vt setting

Assume the metric is Icrit, a first order ADM estimate would be Icrit0/1σIcrit. For example, the SRAM cell of Fig. 5.8 has nominal Icrit0 of 44.4 uA at 0.9 V and 85C. 1 σ Icrit is around 8.7 uA from (5.4). The first order ADM estimate would be 5.1 σ, or stability fail count around 357 ppB. The expected full good stability yield is then around 70% for the 1 M SRAM or 0.3% for the 16 M SRAM. This first order estimate would be fairly accurate if Icrit gradients with respect to Vt skews are fairly constant in the Vt skew space, and the Vt skews are fairly independent of each other.

One way to establish the Vt variations in terms of independent variables is to equate the device Vt σ by Vtmm σ divided by $\sqrt{2}$, where Vtmm is Vt mismatch or the Vt difference between the device pairs in the SRAM cell. This "independent" Vt σ is to be referred as the local Vt σ on a chip or on one wafer. Some compensation for the deviation from the ideal condition of linearity and independence can be made by running circuit simulation to find the critical Vt skew, where the metric M drops below zero. Simulation is done with Vt skews for all 6 FET devices in the "wrong" direction which degrades the metric. The perturbation of the 6 Vt values on SRAM stability is thus realized with the wrong way Vt skews. A unit perturbation vector of the 6 Vt skews is defined which would cause M to drop by 1 σM as defined in (5.4). ADM is then the number of unit perturbations that cause the metric to drop below zero.

$$\delta M = \sum_{i=1}^{6} G_i \ \delta x_i = 1 \ \sigma M = \sqrt{\sum_{i=1}^{6} G_i^2 \sigma_i^2} \tag{5.6}$$

$u_i = \delta \, x_i$ for the unit perturbation that degrades M by 1 σ

For the following discussions, "Vt skew" will refer to the Vt skew in the "wrong" direction that degrades the metric M. In case of Icrit, degrading metric means the decreasing metric. SRAM Icrit decrease with respect to Vt skews is illustrated in Fig. 5.7. G_i can be computed from the Icrit drop at Scrit, the composite Vt skew in mV for all six devices that drops Icrit to 0 from the nominal 44.4 uA for nominal devices of 0 Vt skew In this example, Scrit is around 103.6 mV and the gradient for N0 is then (44.4 uA − 26.7 uA)/103.6 mV. G_i is thus estimated from the average Icrit drop in the range of Vt skew from 0 to Scrit.

There are many unit perturbation vectors $\{u_1, u_2, u_3, u_4, u_5, u_6\}$ that can satisfy (5.6). The most probable worst case unit perturbation vector is the Vt skews that can drop M by 1 σ in the shortest distance

$$u_i = G_i \sigma_i^2 / \sigma M = \delta x_i \quad - \text{most probable vector} \tag{5.7}$$

A simpler perturbation vector is the Murphy vector, where all device Vt skew by the same number of mVs

$$u_i = \sigma M / \sum_{i=1}^{6} G_i \quad - \text{Murphy vector} \tag{5.8}$$

Another simple perturbation vector is the natural vector, where all device Vt skew by the same number of its respective σ.

$$u_i = \sigma_i \left(\sigma M / \sum_{i=1}^{6} G_i \, \sigma_i \right) \quad - \text{natural vector} \tag{5.9}$$

The most probable vector signifies the shortest path of Vt skew to degrade the metric M. The corresponding end point along this shortest path has the highest probability density on the normal distribution of the Vt skews and is hence the most likely failure point in the space of Vt skews. Murphy vector signifies the Vt skew for all devices by the same amount of mVs. The natural vector signifies the Vt skew in proportion to the individual device Vt σ. If all devices have the same Vt σ, all three perturbation vectors would converge to the same vector. Murphy vector is mainly for crude estimate of the SRAM margins when device characteristic is still unknown.

In this formulation, the perturbation δx_i has the voltage unit of Vt skew that would be incompatible with other independent variables in different units which would also affect the operation margin of interest. For example, in the DRM computation, some defect leakage variable in current unit may need to be included for more accurate DRM computation. The computation in multiple variable units becomes very confusing and prone to errors. If the random variable skew is counted in units of its standard deviation, then all relevant random variables would be in units of σ, just as the computed operation margins are. The equations above can then be rewritten in this "normalized" perturbation format.

$$1 \, \sigma M = \sqrt{\sum_{i=1}^{6} G_i^2} \tag{5.10}$$

$$G_i = |\partial M / \partial x_i|$$

$$\delta M = \sum_{i=1}^{6} G_i \, \delta x_i = 1 \, \sigma M = \sqrt{\sum_{i=1}^{6} G_i^2} \tag{5.11}$$

$u_i = \delta \, x_i$ for the unit perturbation that degrades M by 1 σ

x_i is now in units of device and Vt σ instead of mV. G_i is in units of the metric and is free from the units of the perturbation variable. Example of Icrit drop with respect to Vt skews in units of σ is in Fig. 5.33. Normalized unit perturbation vectors are computed in similar manner.

$$u_i = G_i/\sigma M \quad - \text{ most probable vector} \tag{5.12}$$

$$u_i = \sigma M/\sum_{i=1}^{6} G_i \quad - \text{ natural vector} \tag{5.13}$$

As the unit Murphy vector signifies the Vt skew for all devices by the same amount of voltage, the unit natural vector would signify the Vt skew by the same amount of σ.

If Icrit gradients with respect to device Vt skews are constant, it does not matter which perturbation vector is used. Icrit will drop below zero after four units of Vt skews. The margins computed from the different perturbation vectors are equalized by (5.11) or (5.6) which make the different unit perturbations degrade M by the same 1 σM. The 1 σM of (5.10) or (5.4) is still based on most probable vector which is the shortest distance to the boundary of 1 σIcrit drop. The length of the most probable perturbation skew in σ is the ADM. The length of the unit most probable perturbation vector is 1 for 1 σ Icrit drop as illustrated in Fig. 5.5.

Wrong way Vt skews are shown for only two devices of the SRAM inverter holding the "1" state in the Icrit pyramid with idealized straight edges in Fig. 5.4. Icrit in uA declines with upward skew of the PFET P1 |Vt| and with downward Vt

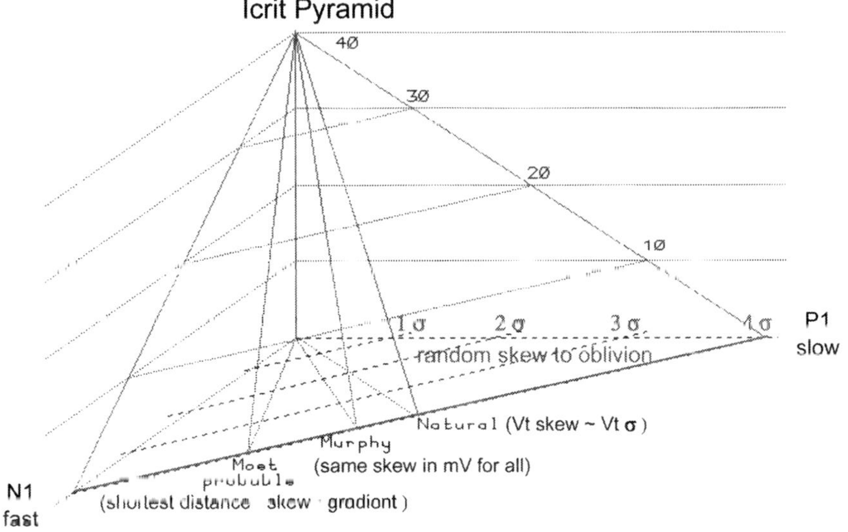

Fig. 5.4 Icrit pyramid to idealize Icrit decline with Vt skews in *straight lines*

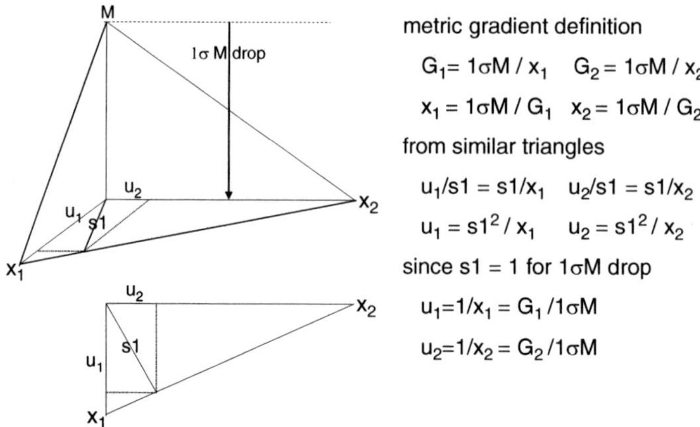

Fig. 5.5 Shortest skew to degrade the metric by 1 σ

skew of the NFET N1. Actual computation will have to include all six devices in the SRAM cell, plus possibly some other independent random variables such as leakage current to and from the cell nodes due to manufacturing defects. Some example real Icrit decline with respect to Vt skews of N1 and P1 are shown in Fig. 5.7.

Equation 5.12 for the most probable perturbation vector represents the device Vt skew along the shortest path in the space of Vt skews to drop the metric by 1 σ. The most probable perturbation path may not look like the shortest path on the bottom plane of the 3D pyramid of Fig. 5.4. It can be seen as the shortest path when viewed on the 2D plane as shown in Fig. 5.5.

Geometrical interpretation is illustrated on the 2D plane of two components of the unit perturbation vector in Fig. 5.5, where u_1 and u_2 are the unit perturbation Vt skews of device 1 and 2.

In case of access disturbs from the word line activation, the total number of perturbation units to flip the cell would be the ADM in units of σ. The corresponding complementary cumulative probability from the normal density function would provide the fail count for one side of the SRAM cell. For example, if the ADM is 5.3 σ, the expected stability fail count for both sides of the SRAM cell is then $2 \times (1 - \Phi(5.3)) = 1.22$ ppM. Here, ppM refers to parts per binary million of $(1024)^2$.

Computation from the most probable perturbation vector or the natural perturbation vector would produce about the same ADM if the Icrit decline is fairly linear with respect to the perturbation skews. The unit perturbation vectors from (5.12) or (5.13) would both satisfy (5.10) and hence would lead to identical ADM if the pyramid edges are perfectly straight. In general, if the Icrit decline with respect to the wrong way Vt skew of the dominating device is concave upward, ADM computed from the most probable vector may be bigger or more optimistic than the ADM computed from the natural vector. More detailed discussion about the two perturbation vectors and the nonlinear Icrit declines is in Sect. 5.13.

5.5 Example ADM and WRM Computation

Stability metric Icrit, Vcrit, or Pcrit from the IV plot, or SNM from the VV plot are all good indicators of SRAM stability. Icrit is chosen in the computation for simplicity and for its linearity with respect to the Vt skews. The decline of the various stability metric from the Vt skew perturbation is illustrated with a typical SRAM cell of the 65 nm node at 0.9 V 85C in Table 5.1 and Fig. 5.6.

Icrit decline and SNM decline with respect to the natural Vt skew perturbation coincide along a straight line in Fig. 5.6. Both metrics would be convenient for ADM computation with straight line approximation. Pcrit and Vcrit are also

Table 5.1 Stability metric decline versus natural Vt skew perturbation in σ. 1 σ natural Vt skew defined by (5.9) or (5.13)

Natural skew (σ)	Stability metrics			
	SNM (mV)	Icrit (uA)	Vcrit (mV)	Pcrit (uW)
0.0	131	44.4	338	8.79
0.5	118	40.2	303	7.35
1.0	106	36.0	272	6.07
1.5	93	31.8	244	4.92
2.0	81	27.6	219	3.90
2.5	69	23.4	196	2.99
3.0	57	19.2	173	2.18
3.5	45	15.0	150	1.49
4.0	32	10.8	125	0.90
4.5	20	6.7	97	0.43
5.0	8	2.5	59	0.10
5.3	0	0.0	0	0.00

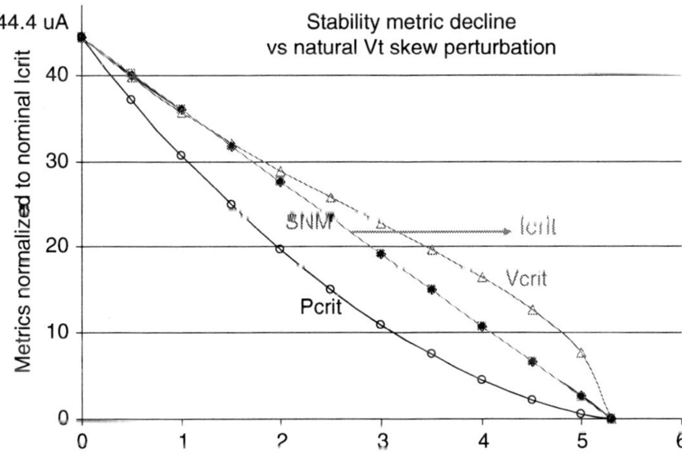

Fig. 5.6 Stability metric versus natural Vt skew perturbation in σ. 1 σ natural Vt skew defined by (5.9) or (5.13)

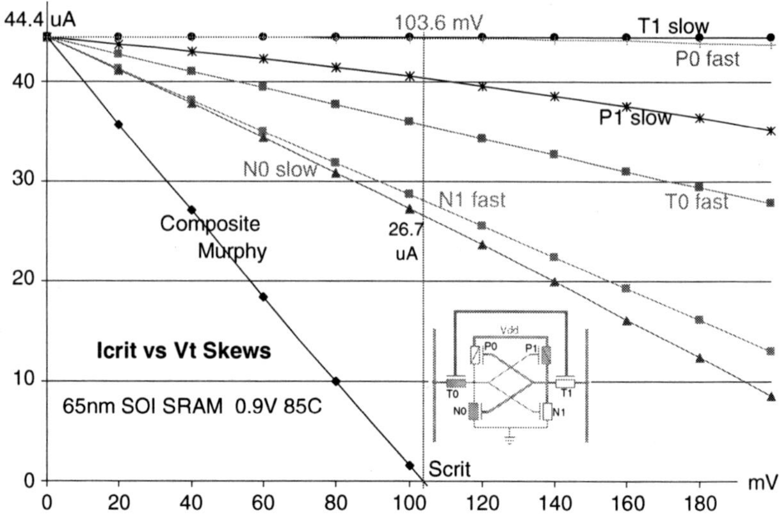

Fig. 5.7 Declining Icrit in uA versus wrong way Vt skews of each and all devices

intuitive indicators of SRAM stability but are not convenient for direct margin computation with linear approximation. In this example, 1 σ or one unit natural Vt skew perturbation is around 0.54 σ Vt skew in the wrong direction for each of the six devices that causes 1 σ Icrit drop of around 8.7 uA. Device Vt σs are around 46 mV, 45 mV, and 34 mV for the passgate pulldown and pullup, respectively in this example.

Icrit gradients with respect to the Vt skew perturbation in mV is illustrated in Fig. 5.7. The SRAM cell would be disturbed when the word line is activated if the natural Vt skew perturbation exceeds 5.3 σ as illustrated in Figs. 5.8 and 5.6. In other words, it takes 5.3 units of the natural perturbation vector to flip the cell during access. This unit natural perturbation is defined by (5.9) or (5.13).

WRM is computed with the Icritw metric under DC write condition in a similar manner as the ADM computation. Icritw gradients w.r.t. the Vt skew perturbation is illustrated in Fig. 5.9. The SRAM cell would not be flipped during the write operation if the natural perturbation approach 7.6 σ as illustrated in Fig. 5.10. This cell has much higher WRM 7.6 σ than ADM 5.3 σ because it is designed for higher performance with relatively fast passgate. SRAM operation yield is thus not limited by WRM but by ADM and RTM.

Computation based on perturbation skews in mV ((5.7)–(5.9)) would produce the same margins as computation based on perturbation skews in σ ((5.12) and (5.13)) if the Icrit gradients with respect to Vt skews are constants. The results may deviate somewhat for the case of nonlinear Icrit drops because the computed gradients may differ. Gradient per device is computed from the nominal Icrit of 0 Vt skew to the Icrit at the critical skew Scrit, where Icrit drops to 0 by the composite Vt skews of all six devices. In the example of Fig. 5.7, the individual

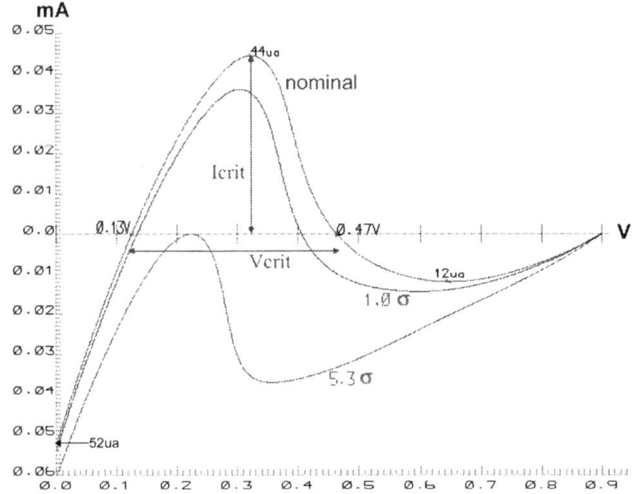

Fig. 5.8 Ncurves of declining Icrit versus wrong way Vt skew. 1 σ natural Vt skew defined by (5.9) or (5.13)

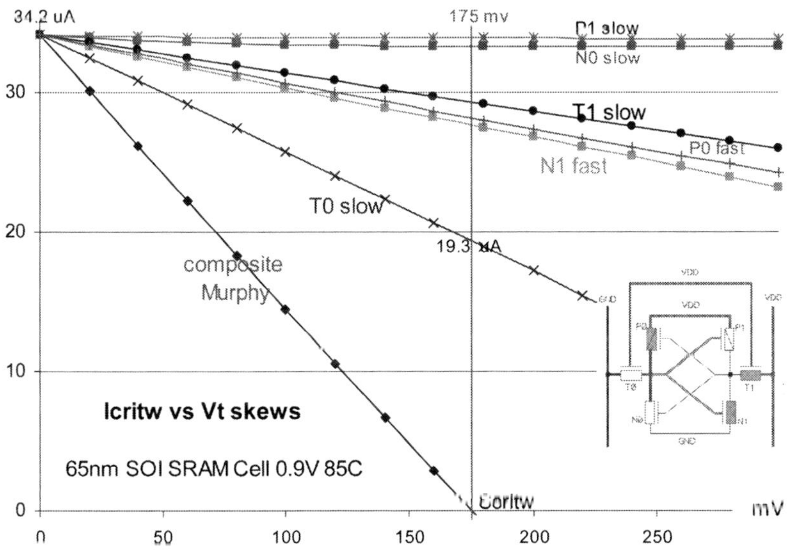

Fig. 5.9 Declining Icritw in uA versus wrong way Vt skews of each and all devices

device G_i is computed at the Scrit of 103.6 mV. The end points at Scrit may be different in the 6D space of Vt skews if the units of Scrit are different. Hence, the computed G_i may differ if Icrit drop is not straight with respect to Vt skews. In this example, the computed ADM with natural perturbation vector from (5.9) is 5.291 σ versus 5.283 σ for the ADM computed from (5.13).

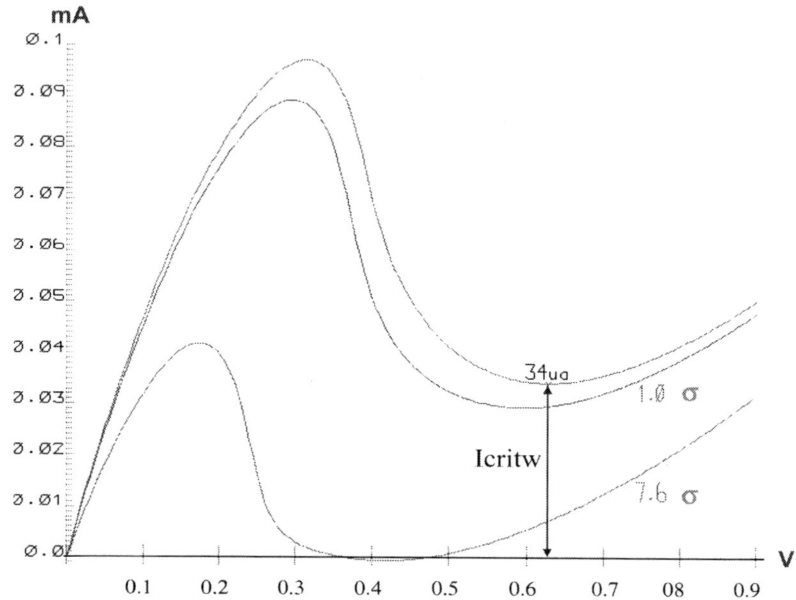

Fig. 5.10 Ncurves of declining Icritw versus wrong way Vt skew. 1 σ natural Vt skew defined by (5.9) or (5.13)

5.6 Extending DC Computation to Transient Operations

The DC operation margins ADM and WRM provide some generic estimate of SRAM fail count under DC conditions. The example ADM above is computed with bit lines and word line clamped to Vdd. Actual access disturb in normal operation is less severe, since bit lines are generally not clamped high during access when word line is pulsed high for limited duration. Transient ADM in actual SRAM environments can be computed with the unit perturbation vectors as defined in the case of DC margin computation.

DC ADM and WRM will be labeled as dcADM and dcWRM to be differentiated from the ADM and WRM for margins in actual transient conditions. For very long bit lines with many cells and for slow operation with long word line pulse width, actual transient ADM would approach the dcADM. Examples of actual transient ADM for a typical 45 nm SOI SRAM are shown in Fig. 5.11. dcADM is from the Icrit computation. ADMO and ADMN are from transient circuit simulation with increasing composite Vt skew until the cell is flipped. The total number of perturbation units is then the ADM in σ.

For SOI devices with floating bodies, ADM would be a function of the body history. ADMO and ADMN are used to specify the stability margins with floating bodies. ADMO represents the "old body" ADM and ADMN represents the "new body" ADM. They are computed with the following operation sequences in consecutive clock cycles.

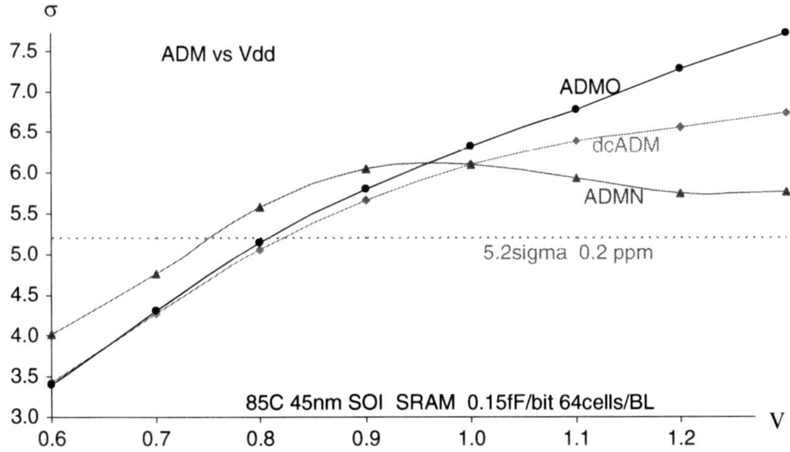

Fig. 5.11 Transient ADM in actual SRAM organization and timing

ADMO: Init0 Access0
ADMN: Init1 Write0 Access0

Here, "0" just signifies one cell state while "1" signifies the opposite state. Init0 or Init1 are to set the cell to a quiescent state, where body levels have settled. Write0 is to activate a word line pulse to write the "0" data from the bit line. Access0 is to activate a word line pulse to the "0" cell not selected for write. The cell is being read or is set to the half-select state, where the word line is selected, but the bit line pair is not selected for read or write. In this example, operating frequency is assumed 1.5 GHz for ADMN and 0.1 GHz for ADMO, and word line pulse width is assumed to be half of cycle time. Simulations are performed across a range of supply voltage Vdd. The cell is counted as a stability fail if it is flipped to "1" during the Access0.

ADMN tends to droop down at high Vdd because device bodies have been bumped to the less stable state in the write cycle. For example, the "on" pulldown device holding the "0" data has body level transiently bumped down by the Write0 operation and become less capable of holding down the "0" node during the immediate subsequent access.

For bulk SRAM with body bias, ADMN and ADMO are the same transient ADM as shown in Fig. 5.12. ADM are not sensitive to the operation history since the device bodies are at constant bias. Bulk ADM is higher because the fixed body bias happens to place the cell in a more stable setting. The difference can be made up by proper Vt tuning in the SOI cell.

Transient WRM of the same cell is shown in Fig. 5.13. Transient WRM is computed as follows

WRMO: Init1 Write0 Access0
WRMN: Init0 Write1 Write0 Access0

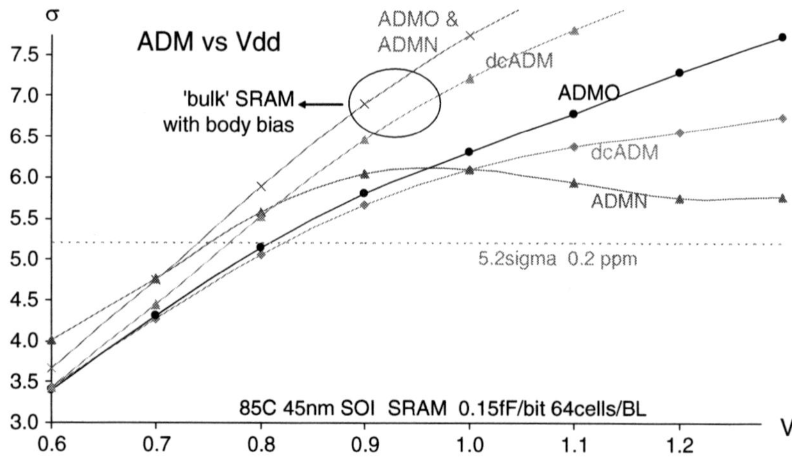

Fig. 5.12 Simpler ADM behavior with body bias as in bulk SRAM

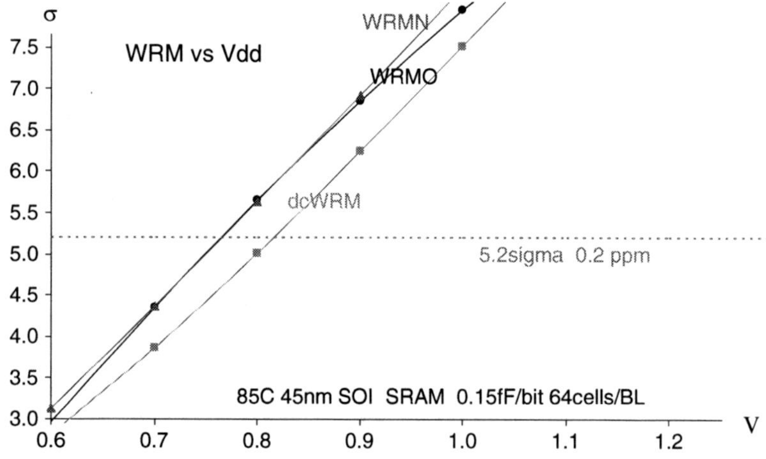

Fig. 5.13 Transient WRM in actual SRAM organization and timing

Once again "0" just signifies one cell state while "1" signifies the opposite state. Operating frequency is assumed 1.5 GHz for WRMO and WRMN, and word line pulse width is assumed half of cycle time across the Vdd range. Write operation is considered a fail if the cell turns out to be in "1" state in Access0.

The transient WRM turns out somewhat higher than dcWRM mainly because the passgate body has been at a higher level since bit lines are at Vdd before the write operation. In dcWRM computation, the "0" bit line is at permanent GND, and so the passgate pulling down the cell node is weaker with the lower body level. For bulk SRAM cells with fixed body bias, transient WRM would be lower than dcWRM as illustrated in Fig. 5.14. dcWRM is computed with word line at

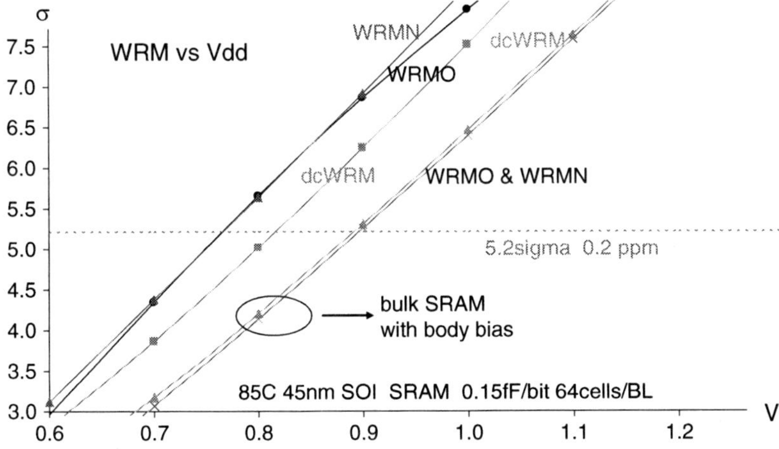

Fig. 5.14 "Bulk" SRAM is less writable but more stable as shown in Fig. 5.12

permanent Vdd while transient WRM is computed with finite word line pulse width that may not be long enough to complete the write operation. The WRM for the same cell with hypothetical body bias are placed next to the WRM of actual cell with floating bodies for comparison.

Bulk WRM is lower than SOI WRM because the fixed body bias happens to place the cell in a more stable setting. ADM becomes higher and WRM becomes lower with the body bias. The differences in ADM or WRM between bulk cell and SOI cell can be reduced with proper device Vt tuning.

Through the rest of the chapter, the default SRAM operating frequency is benchmarked to 1.5 GHz with word line pulse width at 333 pS for transient ADMN and WRM, and 0.1 GHz at word line pulse width at 5 ns for transient ADMO.

5.7 Unstable SRAM Cells at High Vdd and High Beta Ratio

SRAM operation is often specified with Vmin, the lowest Vdd, where SRAM remains functional. ADM and WRM must meet certain application spec at Vdd = Vmin. When scaling reached the 90 nm node, some Vmax problems showed up in the SOI SRAM. Some cells started to get disturbed at high Vdd but became stable again when Vdd was lowered. The high Vdd access disturb may become even more severe when the SRAM β ratio is raised with the intention to improve stability. This unusual behavior is explained with the margin computation in SOI device models.

It can be seen from Fig. 5.11 that the new body ADMN dips when Vdd exceeds 0.9 V. However, there is no Vmax problem if the margin target is 5.2 σ at clock frequency around 1.5 GHz and at bit line loading of 64 cells, since the cell had been tuned with device Vt implants to meet the requirement. Suppose the β ratio is raised

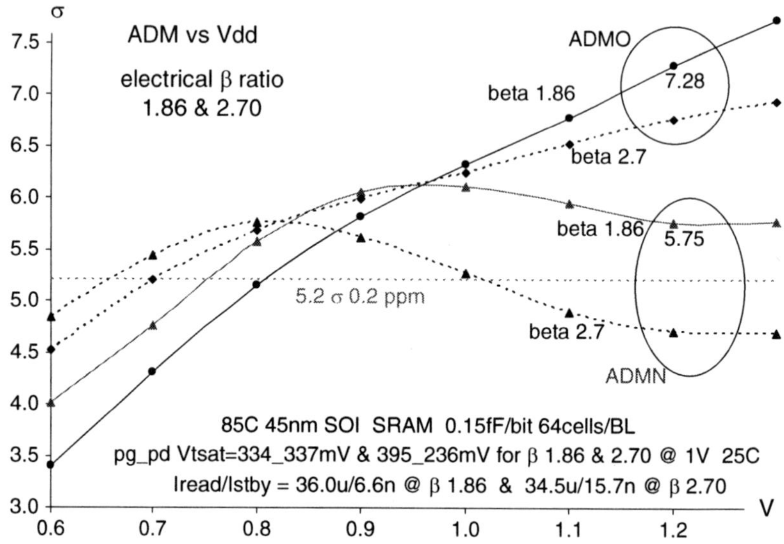

Fig. 5.15 More severe access disturb at higher β ratio 2.70. Faster cooler and more stable at lower β ratio 1.86

with slower passgate and faster pulldown from different Vt implants, the cell would become unstable, in opposite direction of conventional expectation as illustrated in Fig. 5.15.

This stability Vmax problem is mainly due to the new floating bodies which had been bumped from the write operation and are still settling to the quiescent levels as illustrated in Fig. 5.16. In poly gate devices where Tinv is over 1.8 nm, gate to body capacitance is relatively smaller than drain to body capacitance. In a write "0" operation where the left cell node is flipped down, the device floating bodies are bumped in the unstable direction. The pulldown device N0 which was just turned on becomes weaker with lower body level so that β ratio is effectively reduced. The inverter devices P1 and N1 have the body levels just bumped high so that the α ratio is effectively reduced. A lower α ratio means a lower inverter trip point for easier disturb. Bit line restore after write also bumps up the passgate T0 body level which is still settling down. Thus, β ratio is further reduced. The new body cell can end up much less stable than the old body cell. At 1.2 V, ADMN is 5.75 σ versus 7.28 σ of ADMO as shown in Fig. 5.15 for the cell with β ratio 1.86. This ADM droop at high Vdd cannot be compensated with higher β ratio from slower passgate and faster pulldown. The faster pulldown actually makes the ADMN droop at high Vdd more severe since the faster device of lower Vt would be more sensitive to the new floating body levels. At lower Vdd, ADM is higher with higher β ratio since the effects of floating body coupling have diminished.

Body level wave forms are shown in Fig. 5.17 to illustrate the difference between old body disturb and new body disturb at operating frequency of 3.2 GHz and access pulse width around 156 pS. The new bodies are bumped to

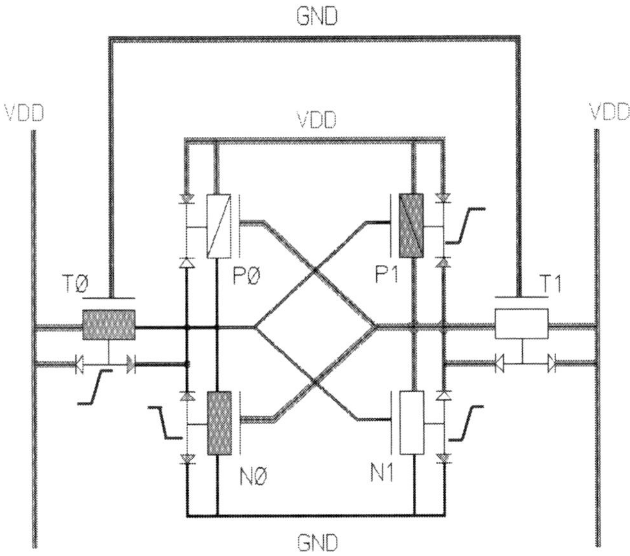

Fig. 5.16 Body bumping from write "0" and bit line restore

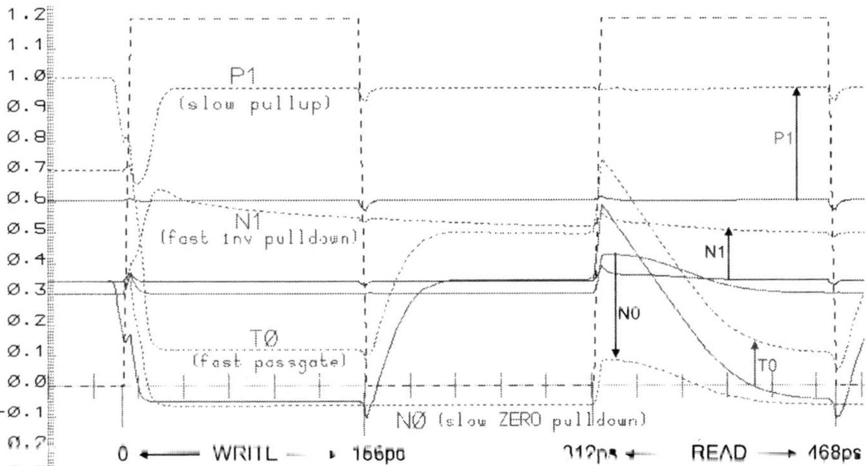

Fig. 5.17 SOI SRAM body levels in a read after write sequence at 3.2 GHz. Old body sequence is Init0 Write0 Read0.

voltage levels that degrade stability by the write operation. It will take a very long time for the new body levels to settle back to the quiescent old body levels that are more favorable for stability.

ADMN vs SRAM cycle time is illustrated in Fig. 5.18. It takes about 100 uS for the floating bodies to settle to quiescent levels. Thus, at frequency below 10 kHz,

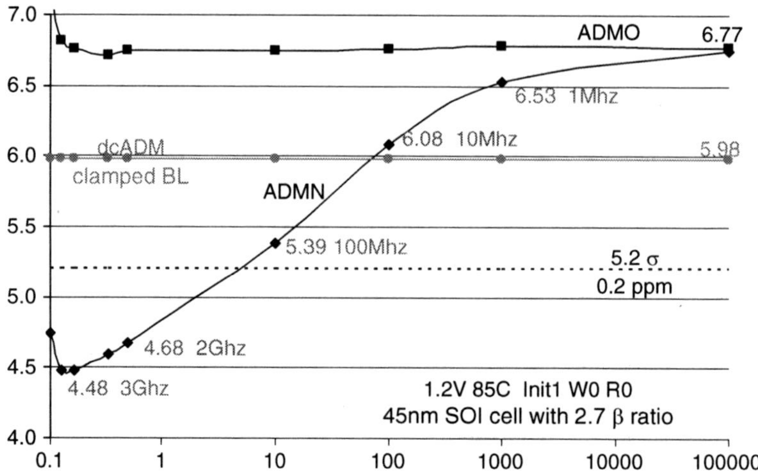

Fig. 5.18 ADM in σ versus cycle time in ns. Word line pulse width of access disturb is half cycle time

floating body effect becomes negligible and ADMN is about the same as ADMO. ADMN drops with higher frequency or shorter cycle time until the cycle time becomes too short. The shortened access pulse around 166 pS at 3 GHz is getting too short to degrade ADMN further.

Thus, for high performance SRAM design at operation frequency over Giga Hertz floating body effects degrading ADMN must be tuned to acceptable levels.

Some ways to reduce the disturbing floating body effects are

(1) Raise device Vt to mitigate the impact of floating bodies
(2) Reduce drain to body capacitance with more leaky junctions
(3) Raise gate to body capacitance with thinner Tinv [1]

More discussion on high Vdd issues is in Sects. 5.10 and 5.12.

5.8 Write Fail from New Floating Bodies

There have been mysterious write fails in SOI SRAM cells, which are designed for fast read/write access. When the γ ratio of Ion_passgate/Ion_pullup is high and the β ratio of Ion_pulldown/Ion_passgate is low, the cell should be easily writable. In all those incidences, the "write" fails turned out to be access disturbs after the write operation when the floating bodies are still settling to the quiescent levels. In actual products, those "write" fails appear erratic in the sense that it may happen sometimes and may not happen some other times. In controlled SRAM tests, these write fails happen more frequently with ripple bit test pattern than ripple word test pattern and the fails simply disappear in DC tests of very long cycle time. Ripple bit pattern refers to the write test of the consecutive bit columns at a fixed word line. Thus, the same word line is activated in every clock cycle until the bits of all columns are completely written. Thus, the cell just written with new data is vulnerable to access disturb while the device bodies are brand new and drifting. In ripple word pattern, the cells along a fixed bit column are written with consecutive word address. The cell written with new data is not exposed to access disturb until the next round of word line ripples. Thus, the body gets more time to settle than the body of the ripple bit case, and hence is less vulnerable to access disturb as shown in Fig. 5.18 which suggests that older bodies provide better stability.

There were also incidences of erratic "write" fails post stress. SRAM designs must pass the stress test before manufacturing to ensure SRAM reliability over the life of the product. Stress test is to run the SRAM in extremely severe environments, such as high Vdd and high temperature, beyond those of the anticipated application conditions. The SRAM stress is expected to slow down the PFETs from NBTI [9, 13] in poly gate cells. Weaker PFETs tend to degrade stability but would enhance writability. However, the write test seems to fail after stress. Again this erratic write fail is due to the access disturb of the SRAM cells that are just written with new floating bodies at fast cycles. Actual culprit of the post stress fail is due to some other defect that had severely reduced the new body ADMN.

Those are the reasons why the WRM is computed with a write operation followed by a subsequent access:

WRMO: Init1 Write0 Access0
WRMN: Init0 Write1 Write0 Access0

A write operation is considered successful only if the cell is not disturbed by the subsequent access. In case of old body write margin WRMO, the cell is flipped over from a very old state. In case of new body write margin WRMN, the cell is flipped over from a brand new state that is just established in a previous write cycle. Actual fail count of the new body disturb in the write test is a very small percentage of total write fails, but they would be included in WRM.

5.9 General Transient Margin Computation Sequence and DC Margin Bias Setting

To capture the various floating body effects, transient margin computation has become very complex and confusing. The general computation sequences are summarized in Table 5.2 for both SOI and Bulk SRAM. ADMO+ and ADMO are identical for poly gate SRAM, where gate-body coupling is negligible relative to drain-body coupling.

Even for high K metal gate device with relative stronger gate-body coupling, the difference between ADMO+ and ADMO is negligibly small. ADMO+ is negligibly bigger than ADMO mainly because the passgate floating body level is slightly

Table 5.2 Operation sequence to measure transient margins

Operation sequence for transient margins of SOI SRAM				
ADMO	Init0	Access0		
ADMO+	Init0	Write0	Access0	
ADMN	Init1	Write0	Access0	
WRMO	Init1	Write0	Access0	
WRMN	Init0	Write1	Write0	Acces0
WRMO+	Init1	Write1	Write0	Acces0
Operation sequence for transient margins of bulk SRAM				
ADM	Init0	Access0		
WRM	Init1	Write0	Access0	

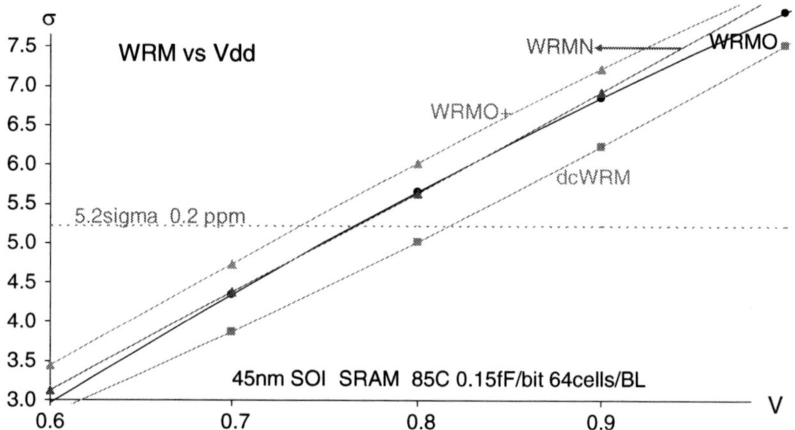

Fig. 5.19 SOI SRAM WRM is defined by the smaller of WRMO and WRMN

lower from the write operation before access. WRMO+ on the other hand is bigger than WRMO and can be ignored in the margin computation. WRMO+ is bigger than WRMO mainly because the Write0 is preceded by Write1 which has raised the floating body level of the passgate which is to pull down the cell node during Write0. This higher WRMO+ is shown in Fig. 5.19 in the background of Fig. 5.13. For bulk SRAM with fixed body bias, transient margins are simply defined by ADM and WRM which are insensitive to the cell body history preceding the access or the write operation.

For completeness, the SRAM bias for DC Margin computation is summarized in Table 5.3. In single supply design, the SRAM terminals are mainly biased to Vdd. Only one bit line is biased to GND for DC WRM computation. The bias would need to be adjusted according to the multiple supply design for various assist features. In Table 5.3, Vcs is cell supply, Vbu and Vbd are the up and down level of bit line. Vwu is the word line up level. For example, bit line down level Vbd may be below GND to assist write; cell supply Vcs may be above Vdd to assist stability; word line up level Vwu may be above Vdd to assist write and below Vdd to assist stability.

5.10   Thinner Tinv to Mitigate Floating Body Effects Which Degrade SRAM Stability

As shown in Fig. 5.16, body bumping by the source switching or drain switching are harmful to SRAM stability, as the β ratio and α ratio are transiently decreased for a long time up to many microseconds. If device electrical gate oxide thickness Tinv [1] is decreased, gate coupling will overwhelm or cancel the effects of body bumping from drain switching. In the following discussion, "metal gate" is to refer to high K metal gate [2] as "poly gate" is referring to poly gate with oxynitride dielectric [3]. The effects of thinner Tinv on stability are illustrated in Figs. 5.20 and 5.21. Device Vtsat is equalized between the poly gate SRAM and the metal gate SRAM for the comparison. Other device parameters with less impact on ADM are left alone.

The metal gate cell has about 23% thinner Tinv than the same cell with poly gate. For a normal design with β ratio 1.92 as shown in Fig. 5.20, the Vmax problem

Table 5.3 SRAM terminal voltage to measure DC margins

	WL	BL	BR	Vcs
SRAM bias for DC margins				
dcADM	Vdd	Vdd	Vdd	Vdd
dcWRM	Vdd	GND	Vdd	Vdd
SRAM bias for DC margins with supply assist				
dcADM	Vwu	Vbu	Vbu	Vcs
dcWRM	Vwu	Vbd	Vbu	Vcs

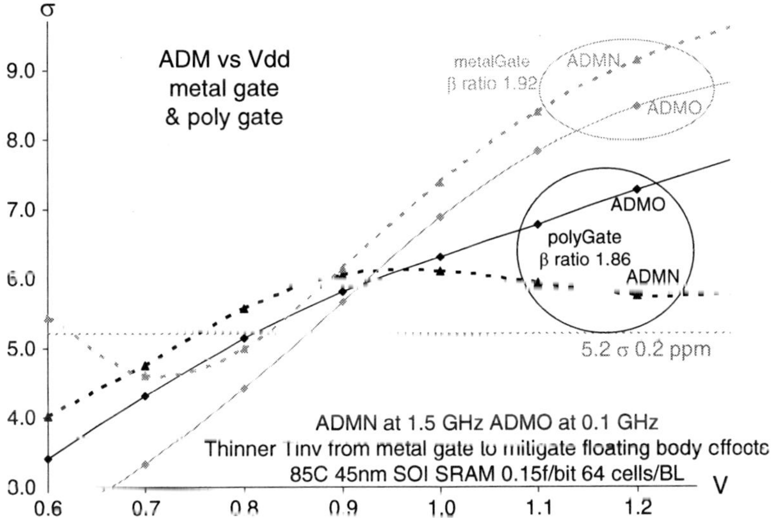

Fig. 5.20 ADM versus Vdd for poly gate and metal gate cell with the same Vtsat. Vtsat = 334_337_238mV for pg_pd_pu @ 1 V 25C

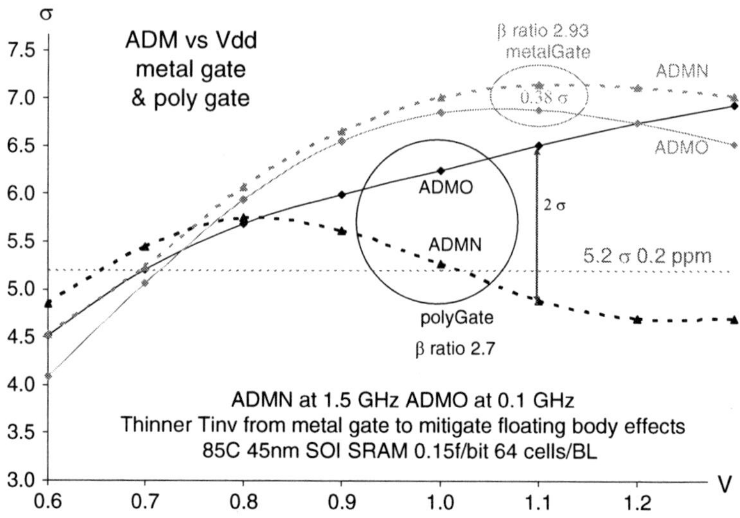

Fig. 5.21 ADM versus Vdd for poly gate and metal gate cell with the same Vtsat. Vtsat = 395_236_238mV for pg_pd_pu @ 1 V 25C

from floating new bodies is completely resolved with the metal gate device. ADMN does not droop down at higher Vdd. The new body ADMN curve is concave upward instead at lower Vdd as expected at high frequency operations. The narrow access disturb pulse around 333 pS at 1.5 GHz is less disturbing to the cell at low Vdd when the cell inverter needs longer time to flip over. When the cell β ratio is raised to 2.93 with faster pulldown, this low Vdd stability bonus is gone since the cell inverter is now fast enough to be flipped over at low Vdd at 1.5 GHz as shown in Fig. 5.21.

Either way, the severe droop of ADMN of the poly gate SRAM is cured by the metal gate, regardless of the β ratios. In Fig. 5.21, margin gap (ADMO–ADMN) drops to −0.38 σ from 2 σ at 1.2 V. Here, Vtsat of the metal gate devices have been adjusted to match Vtsat of the thicker Tinv poly gate devices. As a result of Vtsat matching, device Ion is raised by around 27–39%, and the β ratio is raised to 2.93 from 2.70. In this example, Vt variabiltiy is artificially kept the same to see the net effects of stronger gate to body coupling. Actual metal gate ADM would be higher because device variability would be tightened with the thinner Tinv.

5.11 SRAM Wear and Tear from NBTI and PBTI

SRAM devices are subjected to Vt shift from NBTI and PBTI in the product life [4, 5]. Clock frequency may need to be slowed down to access the SRAM if there is no performance guard band at time 0. Time 0 is referring to the beginning of the product life. NBTI and PBTI also tend to make the cell less stable. Time 0 ADM guard band is thus needed to prevent access disturbs when the SRAM cells become older or near end of product life. Slower PFETs lower the α ratio and the inverter trip point. A lower inverter trip point means easier cell flip by access disturb. PBTI had been negligible in poly gate devices, but is showing up in metal gate devices. PBTI slows down the NFETs and the worn out cells may not be able to deliver the read signal in time. One sided PBTI on one pulldown device is more harmful to stability since the slower pulldown device is less capable of holding down the "0" node. Uniform PBTI on both pulldown devices are less harmful to stability, since the slower "on" pulldown hurts stability but the slower "off" pulldown helps stability. General wear and tear effects on stability are illustrated in Fig. 5.22 and Table 5.4. In poly gate SRAM, only NBTI is of concern because PBTI is relatively negligible. In high K metal gate technology, both NBTI and PBTI would wear and tear the SRAM stability margin which has already been diminished by scaling. To guarantee SRAM functions near the end of life, much bigger design guard bands are needed. If the guard bands are not affordable, some kind of SRAM healing

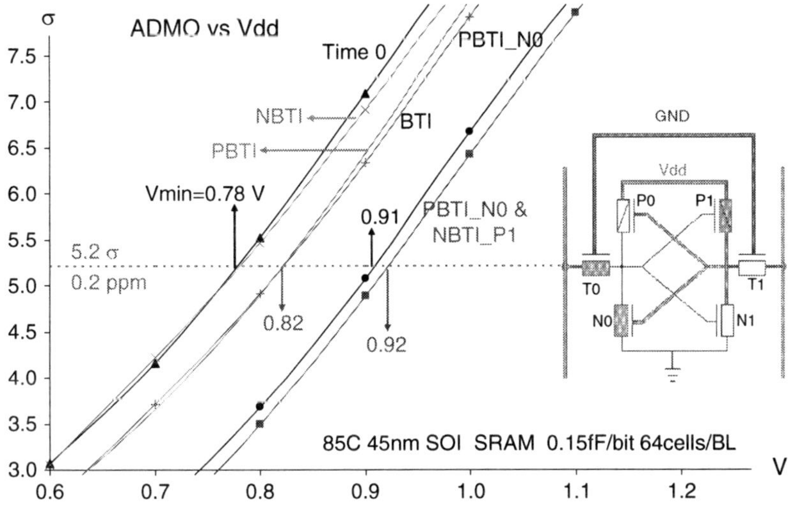

Fig. 5.22 SRAM wear and tear from NBTI and PBTI. Vmin degradation 144 mv in worst case one-sided stress

Table 5.4 Effects of 0.05 V |Vt| shift on stability Vmin at 5.2 σ

	P0	P1	N0	N1	V min	Delta	Comments
Time 0	0	0	0	0	0.776		Brand new SRAM cell
NBTI	0.05	0.05	0	0	0.779	0.002	Nominal PFET |Vt| + 50 mv
PBTI	0	0	0.05	0.05	0.820	0.044	Nominal NFET Vt + 50 mv
BTI	0.05	0.05	0.05	0.05	0.820	0.044	Uniform PBTI & NBTI
wcPBTI	0	0	0.05	0	0.908	0.131	One-sided PBTI
wcBTI	0	0.05	0.05	0	0.921	0.144	One-sided NBTI & PBTI

procedure or application restriction will need to be established for SRAM reliability in the scaled down nodes.

This particular metal gate cell has been tuned for low power operation with long channel and high Vt devices. Iread/Istby of this metal gate cell is around 36 u/0.3 n versus 36 u/6.6 n of the same cell in poly gate. Typical product life Vt shift is around 50 mV. For uniform |Vt| shift of all four devices, stability Vmin degradation is around 44 mV. However, under the rare situation where the cell has been staying in one state in most the product life, Vmin shift will be over 144 mv. This would be strenuous demand on SRAM design to guarantee functionality near product end of life. In this computation, bigger variability from NBTI and PBTI has not been counted. The possible bigger device Vt variability near product end of life will exasperate the Vmin degradation. Only old body stability margin ADMO is shown in Fig. 5.22 because ADMO is generally lower than ADMN when Vdd is under 1 V. This metal gate cell is migrated from the poly gate cell illustrated in Figs. 5.11–5.14. Time 0 operation margins of one migrated metal gate cell are shown in Fig. 5.23.

This migration from the poly gate cell to metal gate cell is for similar Iread at much lower Istby. Iread/Istby is about 36 uA/0.34 nA versus 36 uA/6.6 nA Stability Vmin is defined by old body ADMO, which is very close to dcADM with clamped bit lines. At 64 cells per bit line, the bit line capacitance is big enough to act like a bit line clamp to Vdd. ADMO will coincide with dcADM if bit line loading and access disturb pulse width become infinity. That is also the basis for extending the perturbation vectors from DC metrics to estimate transient margins.

New body ADMN is higher than old body ADMO because the stronger gate to body coupling in metal gate device increases the β ratio and α ratio of the cell with new bodies. ADMN is improved at low Vdd and the ADMN curve becomes concave upward. At 1.5 GHz and around 0.7 V, the relatively slower SRAM cell become more stable since it has less time to flip over from the shorter access disturb pulse. ADMO curve is concave upward around 0.7 V when clock frequency is

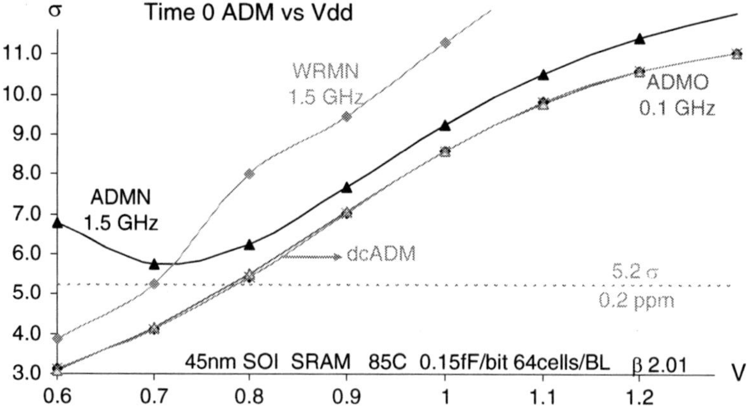

Fig. 5.23 Time 0 SRAM operation margins before wear and tear

raised to 1.5 GHz from 0.1 GHz. This low Vdd effect allows the stability assist with restricted access pulse width. These short-pulse effects help stability even at higher Vdd when the pulse is short enough as shown in Fig. 5.18. When pulse width is below 167 pS at frequency higher than 3 GHz, the cell becomes more stable even at 1.2 V Vdd.

When the devices are tuned to same Vtsats of the poly gate cell for higher Iread at Iread/Istby ratio of 50 u/2.6 n and for more balanced ADM and WRM, wear out effects are shown in Figs. 5.24 and Table 5.5. Worst case ADM Vmin degradation is 111 mV versus 144 mV of the slower Vt setting of Fig. 5.22.

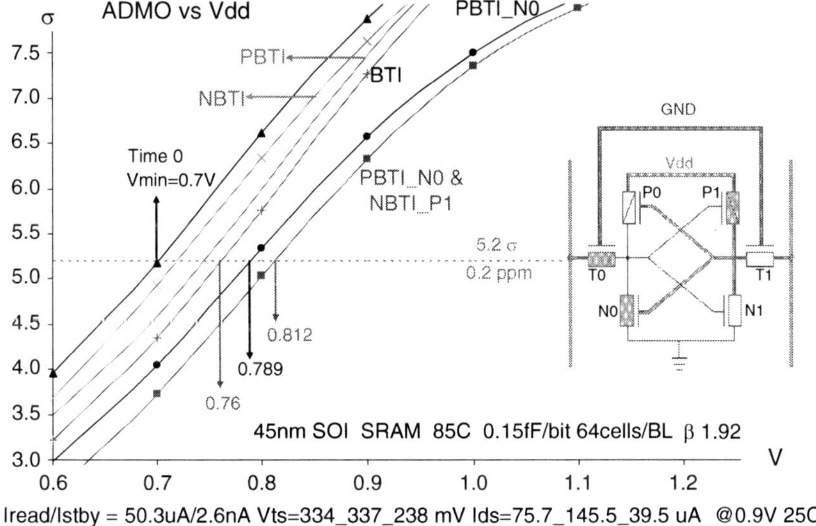

Fig. 5.24 Wear and tear from NBTI and PBTI for faster SRAM

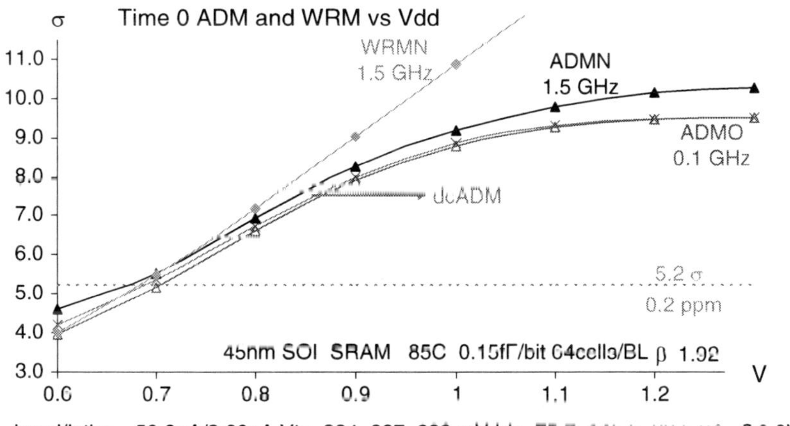

Fig. 5.25 Time 0 operation margins of faster SRAM cell before wear and tear

Time 0 margins are shown in Fig. 5.25. With faster devices, ADMN does not increase with lower Vdd. The pulse width of 333 pS at 1.5 GHz is too wide to assist stability ADMN enhancement at low Vdd by short access pulse only benefits the slower cells as indicated in Fig. 5.23. However the time 0 ADMO at low Vdd is better in the faster cell as shown in Fig. 5.24 in comparison to Figs. 5.22.

As NBTI and PBTI degrade the SRAM stability, SRAM writability may be enhanced since less stable cells tend to be more writable as indicated in Fig. 5.26 and Table 5.6.

Table 5.5 Similar wear out effects of 0.05 V |Vt| shift on faster cell

	P0	P1	N0	N1	Vmin	Delta	Comments
Time 0	0	0	0	0	0.701		Brand new SRAM cell
NBTI	0.05	0.05	0	0	0.719	0.018	Nominal PFET |Vt| +50 mv
PBTI	0	0	0.05	0.05	0.743	0.042	Nomial NFET Vt +50 mv
BTI	0.05	0.05	0.05	0.05	0.760	0.059	Uniform PBTI & NBTI
wcPBTI	0	0	0.05	0	0.789	0.088	One-sided PBTI
wcBTI	0	0.05	0.05	0	0.812	0.111	One-sided NBTI & PBTI

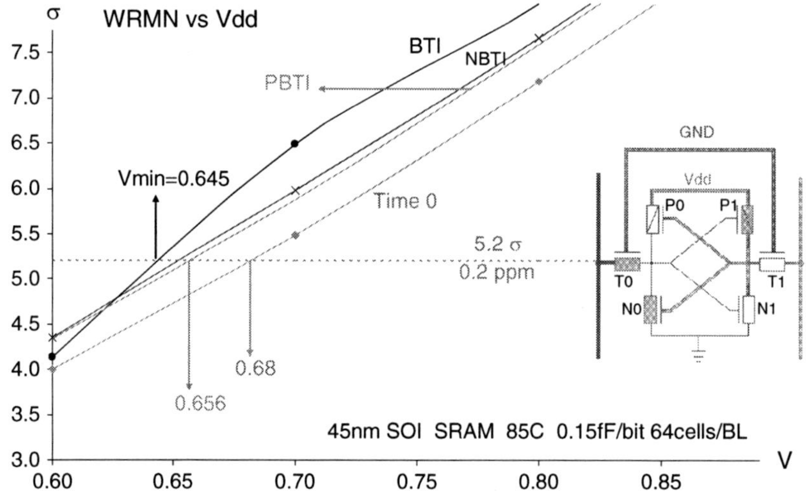

Fig. 5.26 No WRM degradation from NBTI and PBTI. WRM unchanged with one-sided stress that is worst case for ADM drop

Table 5.6 Writability enhancement from NBTI and PBTI

	P0	P1	N0	N1	Vmin	Delta	Comments
BTI	0.05	0.05	0.05	0.05	0.645	−0.035	Uniform PBTI & NBTI
NBTI	0.05	0.05	0	0	0.652	−0.029	Nominal PEFT |Vt| +50 mv
PBTI	0	0	0.05	0.05	0.656	−0.024	Nominal NFET Vt +50 mv
Time 0	0	0	0	0	0.681, 0.680		Brand new SRAM cell
wcBTI	0	0.05	0.05	0	0.681	0.001	One-sided NBTI & PBTI
wcPBTI	0	0	0.05	0	0.682	0.002	One-sided PBTI

5.12   SRAM Vmax Problem from DIBL

Poly gate new body SRAM stability degrades at high Vdd and high frequency as shown in Sect. 5.8. Old body ADMO is generally monotonically increasing with Vdd. Bulk SRAM cell with fixed body bias behaves in a similar manner. This concave downward ADMN curve has been referred to as the "holey" ADM. Acceptable ADM shows up in a hole of some Vdd schmoo plots. ADM stability spec is guaranteed only if Vdd is above a certain Vmin and below a certain Vmax.

Raising the device |Vt| can mitigate the ADM droop at high Vdd but is at the cost of performance. Reducing drain to body capacitance with leakier junctions can help but is at the cost of standby power dissipation. Raising gate to body capacitance with metal gate device appears to be the best remedy since it is necessary anyway for extending CMOS scaling. Thinner Tinv of metal gate SRAM is necessary to reduce device variability to maintain sufficient operation margins in scaled down SRAM cells, as indicated by (5.4) and (5.5), which suggest that operation margins are generally proportional to device Vt σ.

The new SRAM scaling barrier though is the increasing DIBL from the shorter channels beyond 45 nm and 32 nm. Greater DIBL means that the "off" device with short channel may become weakly "on" at high Vdd. That means the "off" device may turn on more easily. Thus, the SRAM cell is more susceptible to access disturb and hence less stable.

These holey ADMs from short channel effects are illustrated in Fig. 5.27 for device Lgate around 30 nm. In this section, Vmin and Vmax are measured by the criterion of 5.8 σ. The cell becomes less stable at higher Vdd, regardless of the body history. In this example, stability Vmin is defined by old body ADMO and Vmax is defined by new body ADMN. ADMN Vmin is assisted by the shorter access pulse width around 333 pS at 1.5 GHz at low Vdd. ADMO is computed for longer access

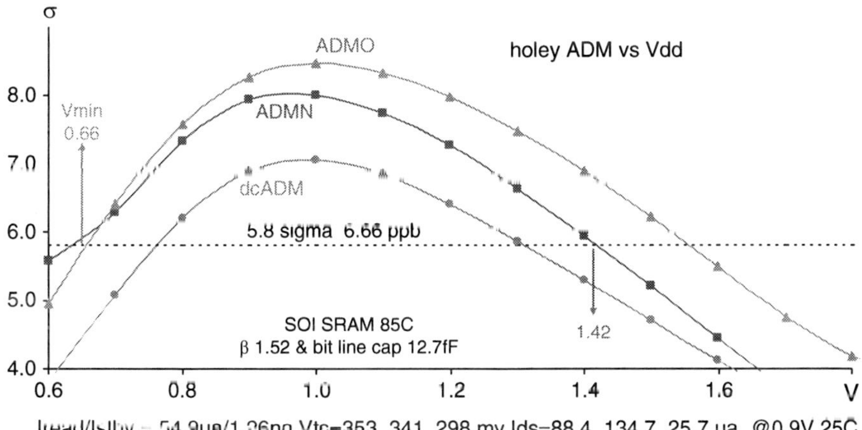

Fig. 5.27 SRAM stability degradation at high Vdd, regardless of body history

pulse width around 5 ns at 0.1 GHz. However, the severe ADM plunge at higher Vdd happens anyway whether the SRAM device bodies are new or old since DIBL happens in all cases. ADMO is bigger than dcADM mainly because bit lines are clamped to Vdd. At 0.1 GHz and with bit line capacitance over 12 fF ADMO would converge to dcADM if the bit lines are clamped as in some older SRAM designs. This ADM droop is more severe for faster SRAM with wider passgate or simply faster NFETs. Passgate width effects are shown in Fig. 5.28.

As the dcADM and ADMO drop at high Vdd for short channel SOI SRAM, it can be inferred that bulk SRAM would face the same Vmax constraint as shown in Fig. 5.29. In bulk SRAM with body bias, ADMN and ADMO are the same ADM.

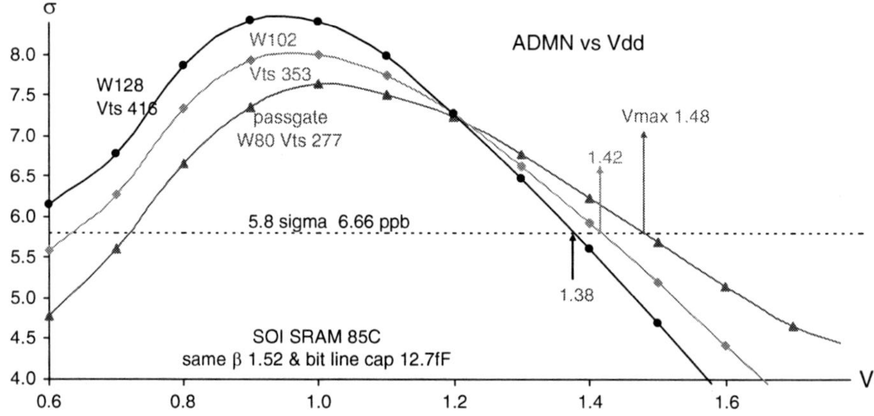

Fig. 5.28 Lower Vmax for wider passgate of tighter Vt variability even though all passgates are tuned to the same β ratio, and the arrays are adjusted to the same total bit line loading. Tighter variability from the wider passgate is also assumed in the simulation

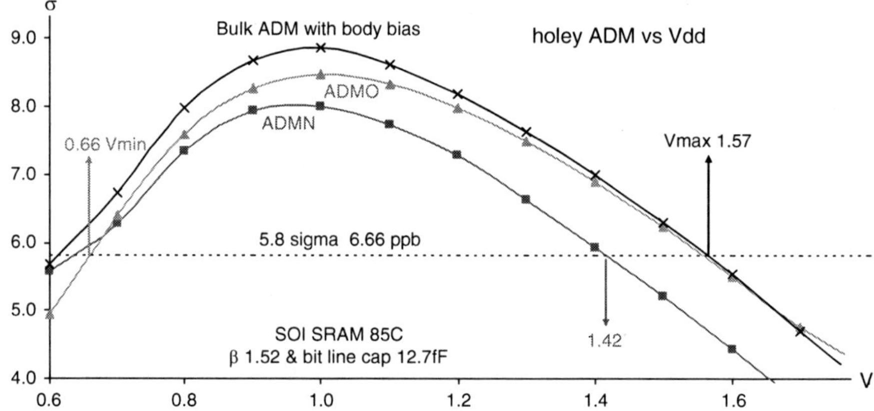

Fig. 5.29 SRAM stability degradation at high Vdd, regardless of body bias

Table 5.7 Effects of 0.05 V |Vt| shift on old body stability

	P0	P1	N0	N1	Vmin	Delta	Comments
NBTI	0.05	0.05	0	0	0.649	−0.009	Nominal PFET |Vt| + 50 mv
Time 0	0	0	0	0	0.658		Brand new SRAM cell
BTI	0.05	0.05	0.05	0.05	0.684	0.026	Uniform PBTI & NBTI
PBTI	0	0	0.05	0.05	0.700	0.042	Nominal NFET Vt +50 mv
wcBTI	0	0.05	0.05	0	0.780	0.122	One-sided NBTI & PBTI
wcPBTI	0	0	0.05	0	0.785	0.127	One-sided PBTI

Table 5.8 Effects of 0.05 V |Vt| shift on new body stability

	P0	P1	N0	N1	Vmax	Delta	Comments
PBTI	0	0	0.05	0.05	1.510	0.091	Nominal NFET Vt +50 mv
BTI	0.05	0.05	0.05	0.05	1.483	0.064	Uniform PBTI & NBTI
Time 0	0	0	0	0	1.419		Brand new SRAM cell
NBTI	0.05	0.05	0	0	1.389	−0.030	Nominal PFET |Vt| +50mv
wcPBTI	0	0	0.05	0	1.287	−0.131	One-sided PBTI
wcBTI	0	0.05	0.05	0	1.249	−0.170	One-sided NBTI & PBTI

Bulk ADM is higher than the SOI ADMO at low Vdd mainly because it is from 1.5 GHz operation while the SOI ADMO is from 0.1 GHz operation. Shorter disturb pulse allows higher stability at low Vdd. "Bulk" cell with body bias had been "tuned" to same Iread and β ratio and device Vtsat as those of the SOI cell.

SRAM wear and tear from PBTI and NBTI will decrease Vmax in addition to raising Vmin. Thus, in scaled down SRAM with worse DIBL than earlier nodes, stability Vmax in addition to Vmin must be tuned to guarantee data integrity in SRAM operation through the product life. Stability Vmin is limited by ADMO at slow cycles since it cannot benefit from the "virtual" assist from narrow access pulse. ADMO Vmin push up from PBTI and NBTI is illustrated in Fig. 5.30 and Table 5.7. The one-sided stress on the SRAM would push up Vmin 127 mV, somewhat similar to the earlier cases of longer Lgate. Vmax pull down of 170 mV is illustrated in Fig. 5.31 and Table 5.8. Thus, the stable Vdd operating range for a brand new cell is [0.66, 1.42] as defined in Fig. 5.27 may be shortened to [0.8, 1.25] near end of life as indicated in Fig. 5.30 and Fig. 5.31. Time 0 stability Vmin 660 mV is defined by ADMO. Post stress Vmin 802 mV is defined by ADMN. Time 0 Vmax and post stress Vmax are both defined by ADMN.

5.13 Getting Around Curvatures of Metric Gradients

Transient Margin computation of bulk SRAM cell with body bias is much simpler without the complexity of floating body effects. Complexity comes at lower Vdd for the low power SRAM cells with Vtlin around Vdd. When Vdd-Vtlin is around

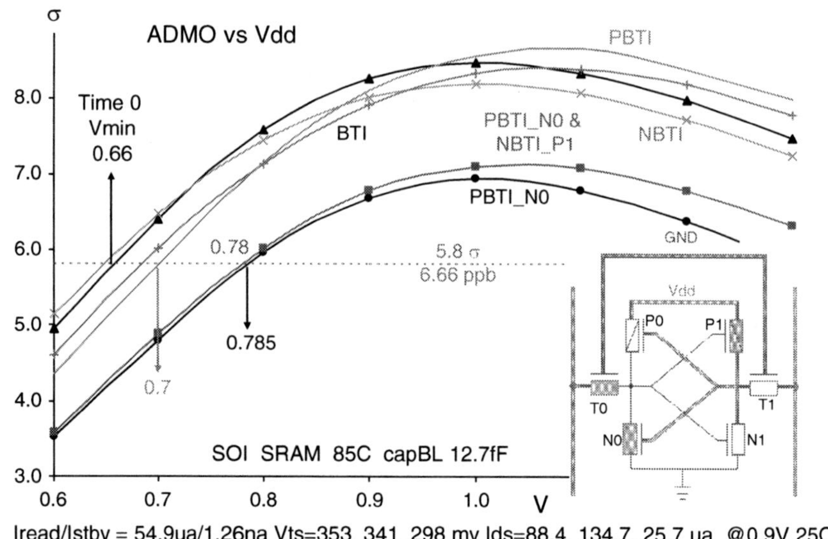

Fig. 5.30 Wear and tear from NBTI and PBTI of ADMO at 0.1 GHz

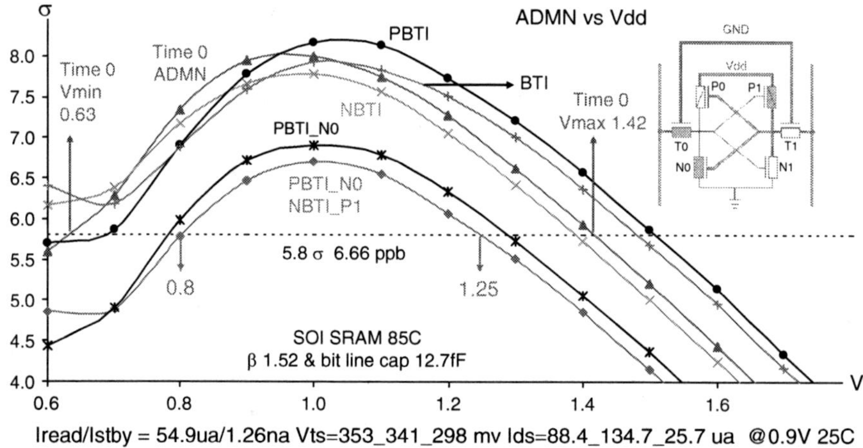

Fig. 5.31 SRAM stability degradation at high Vdd, regardless of body bias. Time 0 Vmin defined by ADMO is 0.66 V

zero, device is mainly operating in the linear region. The Icrit gradient versus device Vt skew perturbation may deviate quite far off the straight line approximation as illustrated in Fig. 5.32. As N0 is perturbed deeper into the subthreshold region with higher Vt, Icrit curve with respect to Vt skew is concave upward since the effect of further Vt skew is diminishing. As T0 is perturbed away from the subthreshold region with lower Vt, Icrit curve with respect to Vt skew is concave

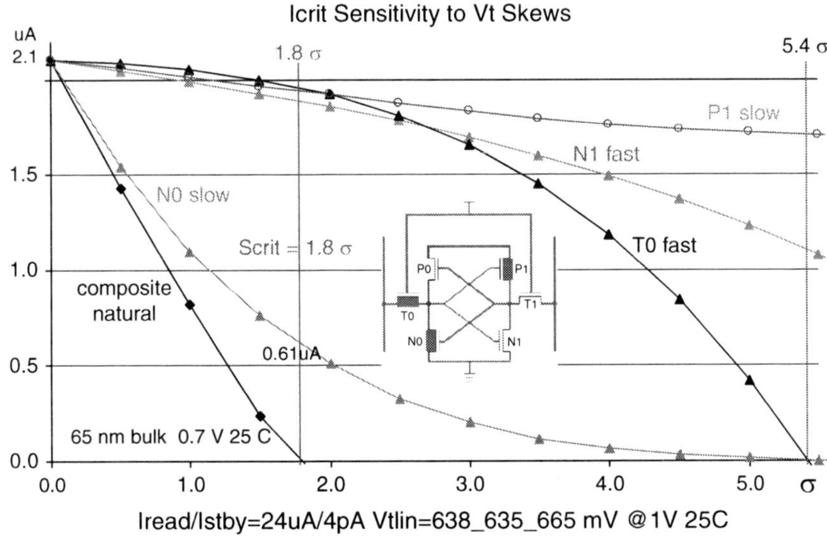

Fig. 5.32 Nonlinear Icrit gradients of low power SRAM at low Vdd. Composite natural Icrit drop from Vt skews of all devices

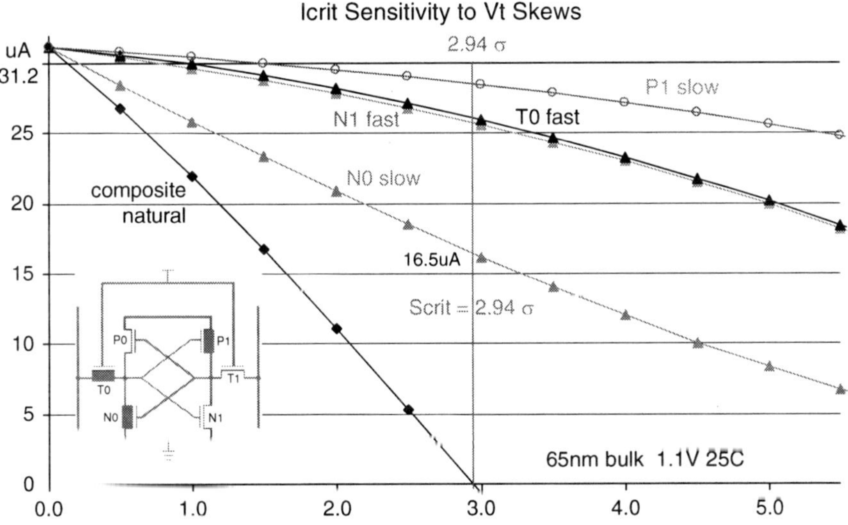

Fig. 5.33 Declining Icrit in uA versus wrong way Vt skew in σ. Composite natural Icrit drop from natural Vt skews of all devices. More linear gradients at 1.1 V for same cell of Fig. 5.32

downward as the effect of further Vt skew is increasing. The milder gradient deviation from straight line at higher Vdd at 1.1 V is shown in Fig. 5.33 for comparison. Icrit drop with respect to Vt skews of P0 and T1 is about 0, and

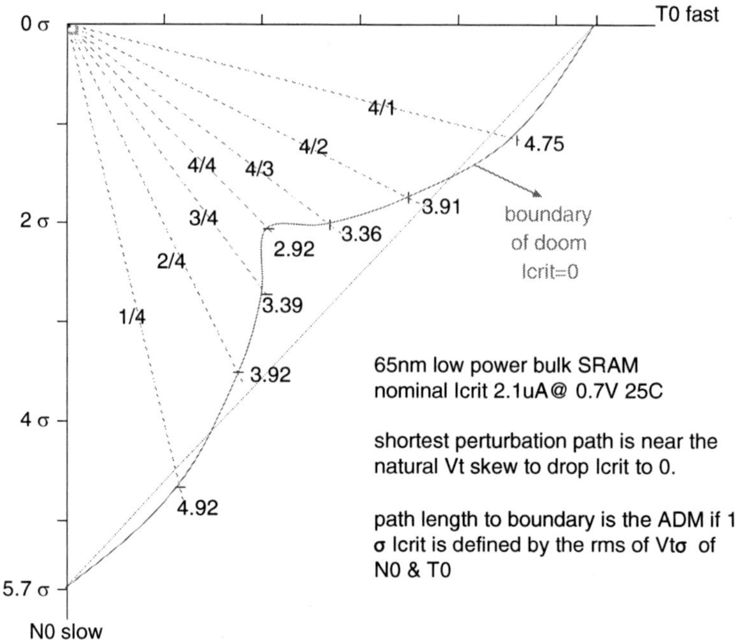

Fig. 5.34 Vt skew paths of N0 and T0 to drop Icrit to 0

hence not shown in the chart. Standard FET symbols are to signify bulk devices. The nonstandard symbols for SOI cells in earlier pictures are to signify SOI devices.

The random Vt skews of N0 and T0 are sketched in Fig. 5.34. Among the various perturbation paths to flip the SRAM cell, the natural path with wrong way Vt skew in proportion to its individual Vt σ of N0 and T0 is the worst case perturbation path. It is worst case in the sense that the corresponding normal density function would predict the maximum fail counts. Nominal Icrit of 2.1 uA occurs at the left upper origin where all devices are at its nominal Vt. Other than N0 and T0, all other four devices are assumed to stay nominal without any perturbations in this random skew exercise of N0 and T0. Actual margin computation becomes very complicated. Since the gradients are changing along the perturbation path, the 1 σIcrit as defined by (5.4) and (5.10) becomes a variable along the way to oblivion.

Icrit gradients are calculated at the critical natural Vt skew Scrit of 2.94 σ as shown in Fig. 5.33 for the case of 0.9V Vdd. Scrit refers to the Vt skew for all devices that drops the SRAM Icrit to 0. For example, the N0 gradient for positive Vt skew to Scrit is (31.2 uA $-$ 16.5 uA)/2.94 σ. When Vdd $=$ 0.7 V which is close to the device Vtlin, this Scrit drops to 1.8 σ as shown in Fig. 5.32. At this low Vdd, the Icrit gradient computed in this manner is too steep for N0 and too shallow for T0. The ADM computed from natural perturbation vector would be too pessimistic and the ADM computation from most probable perturbation is too optimistic as can be inferred from Fig. 5.34. As the N0 skew is dominantly weighted with the steep Icrit gradient, the composite natural skew is very close to the N0 skew axis. The Icrit

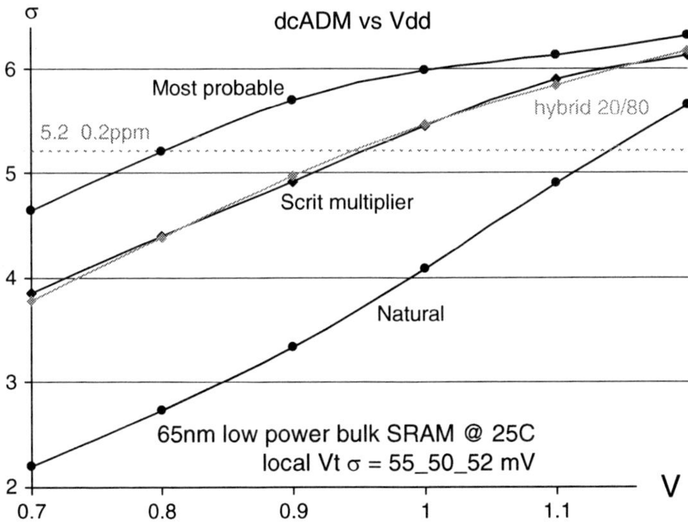

Fig. 5.35 dcADM empirical fitting with hybrid perturbation and with Scrit end point extension

drops abruptly at the beginning, but the decline slows down, and Icrit becomes almost flat near the boundary of doom, where Icrit drops to 0 and the cell data is lost. As the Icrit curves with respect to Vt skews deviate from straight lines, so does the boundary of doom in the space of Vt skew. However, using a gradient estimate with the Scrit end point projects a much shorter walk to the boundary of doom. The net is that the most probable perturbation would lead to higher and more optimistic ADM while the natural perturbation would lead to lower and more pessimistic ADM as indicated in Fig. 5.35.

A quick patch for this deviation is to mix the natural vector and the most probable vector in various proportions and pick the one that fits hardware data. This is referred as the hybrid perturbation vector and some empirical proportion to fit the hardware is 20% and 80% for the natural and the most probable perturbation. The hybrid perturbation vector is defined by (5.14).

$$u_i\text{\_hybrid} = pFactor \times u_i\text{\_natural} + oFactor \times u_i\text{\_mostProbable} \qquad (5.14)$$

pFactor is the pessimistic factor, e.g. 20%
oFactor is the optimistic factor, e.g. 80%
u_i_natural is unit perturbation defined by (5.13)
u_i_mostProbable is the unit perturbation defined by (5.12)

Since the unit natural vector defined by (5.13) and the unit most probable vector defined by (5.12) satisfy (5.11), the unit hybrid vector defined by (5.14) will also satisfy (5.11). The composite Vt skew of the six devices according

Table 5.9 ADM fitting with modified unit vectors from modified Scrit

Vdd (V)	Natural dcADM (σ) no patch	Fitting Scrit multiplier	Natural dcADM (σ) with patch	ADM patching ratio
0.6	1.79	1.68	3.37	1.88
0.7	2.21	1.58	3.85	1.75
0.8	2.73	1.48	4.40	1.61
0.9	3.34	1.38	4.92	1.47
1.0	4.08	1.28	5.45	1.33
1.1	4.90	1.18	5.90	1.20
1.2	5.65	1.08	6.13	1.09

Scrit is the critical composite Vt skew that drops Icrit to 0

to (5.14) will cause 1 σ drop of the metric M in the idealized case of constant metric gradients.

An alternative quick patch is to raise the natural ADM by computing Icrit gradients at Vt skew further away from Scrit, which is the critical composite natural Vt skew that drops Icrit to 0. Icrit gradients have been computed with (Icrit_nom – Icrit@Scrit)/Scrit. Icrit_nom is the nominal Icrit of 0 Vt skews. The patch is to compute Icrit gradients with (Icrit_nom – Icrit@Scritm)/Scritm, where Scritm defines the modified end point to compute Icrit gradients. This patch is illustrated on Table 5.9, where Scritm is acquired from Scrit with some fitting multiplier. Column 2 has the pessimistic ADM from natural vector computation with gradients computed at Scrit. Column 3 has the fitting Scrit end point multipliers to compute ADM of column 4 with gradients computed at Scritm to fit hardware data. Essentially, the Scrit multipliers change the Icrit gradients which would modify 1 σM and the unit vectors that lead to a bigger ADM. The fitting Scrit multiplier (fsm) is a good indicator of how far the real margin deviates from the idealized margin with linear approximation. In this exercise, the fsm gradient with respect to Vdd happens to be 1.

These empirical patches on ADM are shown in Fig. 5.35, where hardware data are around the ADM curves from hybrid vector or from the modified natural vector with Scrit end point multiplier.

The ADM curves are from model-hardware correlation exercise with preliminary device models and early hardware. These corrective patches for nonlinear gradients are empirical. They can only serve for better estimates of SRAM margins in new technologies until some more precise methodology is developed. One possible refinement is to apply piece-wise computation on small intervals of Vt skews.

It can be seen from Fig. 5.35 that the Vmin for this SRAM cell needs to be above 0.95 V for some workable stability yield with dcADM above 5.2 σ. Actual transient ADM is very close to dcADM when the bit line loading is over 256 cells, and the access pulse width is over 5 ns as in the typical low power SRAM applications. The large device Vt σ in this early development phase has to come down through process yield learning to achieve better Vmin and higher yield.

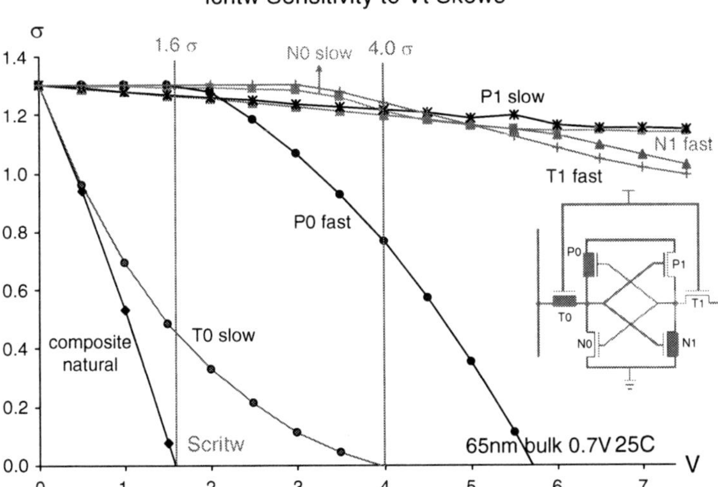

Fig. 5.36 Declining Icritw in uA versus wrong way Vt skew in σ

Write margin computation of low power cell at low Vdd is complicated in a similar manner. Icritw gradients for the same cell at 0.7 V are shown in Fig. 5.36. As T0 is skewed deeper into the subthreshold region, Icritw curve is concave upward since the effect of Vt skew is diminishing. As P0 is skewed away from the subthreshold region, Icritw curve is concave downward since the effect of Vt skew is increasing. Corrective patches for nonlinear gradients on WRM would be needed as in the case of ADM.

5.14   Conclusion and Acknowledgment

Expected fail counts in SRAM operations have been represented in operation margins in units of standard deviations. Direct computation with perturbation of device characteristics has been described in some details for ADM and WRM of standard 6 T SRAM cells

ADM – Access Disturb Margin for the expected stability fail counts
WRM – WRite Margin for the expected write fail counts

Some other SRAM operation margins can be computed in a similar manner.

DRM – Data Retention Margin for the expected data loss
RTM – Read Time Margin for the expected fail counts of read access

In the case of DRM, the suitable metrics are Icritu and Icritd of Fig. 5.1 for the standby cells on inactive word lines. One possible additional random variable is leakage current from process defects or radiations which are independent of device characteristics. DRM is important to indicate SRAM data integrity in standby mode or power down mode and for general soft error rate estimate. In case of RTM, a suitable metric is the magnitude of the bit line signal at the falling edge of the word line pulse. Suppose the sense circuitry is designed to sense a 100 mv bit line signal, the RTM fail criterion is this 100 mV. When device |Vt| is skewed up, the sense signal will drop until the bit line signal is less than 100 mV when the access pulse is going away. These "wrong" way Vt skews are then used for the RTM computation. The 6 T SRAM computation scheme has also been extended to include more than 6 Ts for multiport SRAM cells and for aggressive support circuitry constraining the SRAM operation margins.

The issues of random device Vt skews appear to be simple subsets of the random walk and ruin problems in probability theory [6]. In the symmetric random walk, the gambler has equal chance of winning and losing in a fair bet. In this margin computation, the device Vt skews are "guided" to take the "asymmetric" random walk only in the "wrong" direction with no chance of winning. The Murphy perturbation vector of (5.8) is named after Murphy's adage that states "Anything that can go wrong will go wrong". The natural perturbation vector of (5.9) or (5.13) is actually the "natural" Murphy vector which is in units of σ instead of mV. It is just natural that device Vt would skew according to its σ. Unlike Wald's theory of sequential sampling with random walk in arbitrary numbers of unit steps [14], the random walk of the device Vt skews are counted in unit perturbation that degrades the metric by 1 σ.

This methodology has been started in IBM around the time between 130 nm node and 90 nm node when SRAM scaling was starting to get difficult. It has also been used and adjusted for SRAM margin estimates through SRAM developments of the multiple scaled down generations. The IBM management and many colleagues in the Fishkill semiconductor technology alliance have provided valuable support, feedback, and suggestions. In particular, Pamela Castalino and Mustafa Akbulut had developed a very elegant graphic user interface that looks like a simple calculator to get SRAM operation margins at the finger tips. Pamela had also ported the programs in IBM PowerSPICE to other commercial circuit simulators for the alliance common platform. The computation and terminology has been mainly used inside IBM and alliance and not well publicized outside. Some publications partially relevant to this chapter are in references [7–11].

The SRAM Ncurves or IV plots had been the common IBM stability monitors in the ancient bipolar era when cell data was clamped to only one base-emitter voltage (Vbe) around 0.7 V. They are brought back to the foreground to help analysis of SRAM scaling that has severely degraded operation margins at scaled down Vdd around or below the bipolar Vbe. Icrit or Icritw from the IV plot has been used in this chapter for its simplicity in computation and measurement. Common SRAM leakage problems can be directly interpreted from the IV plot of the bad cell. Multiple Ncurve charts are easier to analyze than multiple butterfly charts, where two VV curves have to be correctly paired up among the many VV curves.

Appendix: SRAM Margins to Meet Yield Targets

All good chip yield (%)	1 M chip		16 M chip		32 M chip		64 M chip	
	Fail count (ppM)	Margin (σ)	Fail count (ppM)	Margin (σ)	Fail count (ppM)	Margin (σ)	Fail count (ppM)	Margin (σ)
1	4.3918	4.59	0.2745	5.14	0.1372	5.27	0.0686	5.39
5	2.8570	4.68	0.1786	5.22	0.0893	5.35	0.0446	5.47
10	2.1959	4.73	0.1372	5.27	0.0686	5.39	0.0343	5.52
20	1.5349	4.81	0.0959	5.33	0.0480	5.46	0.0240	5.58
30	1.1482	4.86	0.0718	5.39	0.0359	5.51	0.0179	5.63
40	0.8738	4.92	0.0546	5.44	0.0273	5.56	0.0137	5.68
50	0.6610	4.97	0.0413	5.49	0.0207	5.61	0.0103	5.73
60	0.4872	5.03	0.0304	5.54	0.0152	5.66	7.61E-03	5.78
70	0.3402	5.10	0.0213	5.60	0.0106	5.72	5.31E-03	5.84
80	0.2128	5.19	0.0133	5.68	6.65E-03	5.80	3.33E-03	5.91
90	0.1005	5.33	0.0063	5.81	3.14E-03	5.92	1.57E-03	6.04
99	0.0096	5.74	5.99E-04	6.19	3.00E-04	6.30	1.50E-04	6.41
99.99	9.537E-05	6.47	5.96E-06	6.88	2.98E-06	6.98	1.49E-06	7.08

Fail count is from both sides of the normal density function beyond margin σ

Grade A > 6.0 σ < 2.0 ppB – parts per binary billion of $(1024)^3$

Grade B > 5.2 σ < 0.2 ppM – parts per binary million of $(1024)^2$

Grade C > 4.9 σ < 1.0 ppM

References

1. Lo S-H, Buchanan DA, Taur Y, Wang W (1997) Quantum-mechanical modeling of electron tunneling current from the inversion layer of ultra-thin-oxide nMOSFET's. IEEE Electron Device Lett 18(5):209–211
2. Mistry K, Allen C, Auth C, Beattie B, Bergstrom D, Bost M, Brazier M, Buehler M, Cappellani A, Chau R, Choi C-H, Ding G, Fischer K, Ghani T, Grover R, Han W, Hanken D, Hattendorf M, He J, Hicks J, Heussner R, Ingerly D, Jain P, James R, Jong L, Joshi S, Kenyon C, Kuhn K, Lee K, Liu H, Maiz J, McIntyre B, Moon P, Neirynck J, Pae S, Parker C, Parsons D, Prasad C, Pipes L, Prince M, Ranade P, Reynolds T, Sandford J, Shifren L, Sebastian J, Seiple J, Simon D, Sivakumar S, Smith P, Thomas C, Troeger T, Vandervoorn P, Williams S, Zawadzki K (2007) A 45 nm logic technology with high-k + metal gate transistors, strained silicon, 9 Cu interconnect layers, 193 nm dry patterning, and 100% Pb-free packaging. IEDM Tech Dig, 247–250
3. Kim YO, Manchanda L, Weber GR (1995) Oxynitride-dioxide composite gate dielectric process for MOS manufacture, US Patent Number 5,464,783, Nov. 7, 1995
4. Pae S, Agostinelli M, Brazier M, Chau R, Dewey G, Ghani T, Hattendorf M, Hicks J, Kavalieros J, Kuhn K, Kuhn M, Maiz J, Metz M, Mistry K, Prasad C, Ramey S, Roskowski A, Sandford J, Thomas C, Thomas I, Wiegand C, Wiedemer J (2008) BTI reliability of 45 nm high-K metal-gate process technology. Annual international reliability physics symposium proceedings, 2008, pp 352 357
5. Drapatz S, Georgakos G, Schmitt-Landsiedel D (2009) Impact of negative and positive bias temperature stress on 6 T-SRAM cells. Adv Radio Sci 7:191–196

6. Feller W (1965) An introduction to probability theory and its applications. In: Random walk and ruin problems, Chapter 14. vol 1, 2nd edn, Wiley
7. Wann C, Wong R, Frank D, Mann R, Ko S-B, Croce P, Lea D, Hoyniak D, Lee Y-M, Toomey J, Weybright M, Sudijono J (2005) SRAM cell design for stability methodology, 2005 IEEE VLSI-TSA international symposium on VLSI technology, pp 21–22
8. Rosa G.L., Loon NW, Rauch S, Wong R, Sudijono J (2006) Impact of NBTI induced statistical variation to SRAM cell stability. Annual international reliability physics symposium proceedings, 2006, pp 274–282
9. Wang L, Ye Q, Wong R, Liehr M (2007) Product burn-in stress impacts on SRAM array performance. International reliability physics symposium proceedings, 2007, pp 666–667
10. Yang S, Wong R, Hasumi R, Gao Y, Kim NS, Lee DH, Badrudduza S, Nair D, Ostermayr M, Kang H, Zhuang H, Li J, Kang L, Chen X, Thean A, Arnaud F, Zhuang L, Schiller C, Sun DP, The YW, Wallner J, Takasu Y, Stein K, Samavedam S, Jaeger D, Baiocco CV, Sherony M, Khare M, Lage C, Pape J, Sudijono J, Steegan AL, Stiffler S (2008), Scaling of 32 nm Low Power SRAM with High-K Metal Gate. IEEE International on Electron Devices Meeting, IEDM technical digest, 2008, pp 233–236
11. Bauer F, Georgakos G, Schmitt-Landsiedel D (2009) A design space comparison of 6T and 8T SRAM core-cells. Integrated circuit and system design. Power and timing modeling, optimization and simulation, Springer, pp 116–125
12. Seevinck E, List F, Lohstroh J (1987) Static noise margin analysis of MOS SRAM cells. IEEE J Solid-State Circuits 22(5):748–754
13. Schroder DK (2007) Negative bias temperature instability: what do we understand? Microelectron Reliab 47:841–852
14. Wald A (1944) On cumulative sums of random variables. Ann Math Stat 15:283–296

Chapter 6
Yield Estimation by Computing Probabilistic Hypervolumes

Chenjie Gu and Jaijeet Roychowdhury

6.1 Introduction: Parameter Variations and Yield Estimation

Parameter variations are inevitable in any IC process. Process steps such as oxidation, doping, molecular beam epitaxy, etc., are all fundamentally statistical in nature [1]. Design of functioning circuits and systems has traditionally relied heavily on the presumption that the law of large numbers [2] applies and that statistical averaging predominates over random variations – more precisely, that the statistical distributions of important process, geometrical, environmental, and electrical parameters cluster closely about their means. Unfortunately, with feature sizes having shrunk from 90 to 65 nm recently (with further scaling down to 45 and 32 nm predicted by the ITRS roadmap [3]), this assumption is no longer valid – in spite of efforts to control them [4, 5], large variations in process parameters are the norm today. This problem is most severe for circuits that try to use the minimum transistor size (e.g., memory circuits [6] for which chip area is of high priority). With transistors having become extremely small (e.g.: gates are only 10 molecules thick; minority dopants in the channel number in the 10s of atoms), small absolute variations in previous processes have become large relative ones. Lithography-related variability at nanoscales [5], which affect geometrical parameters such as effective length and width, further compound parameter variation problems.

Due to such variations in physical, process or environmental parameters, electrical characteristics of circuit devices (e.g., MOS threshold voltages) change; this leads to undesired changes in circuit performances. For example, in static RAM (SRAM) circuits, like the 6T cell [7] in Fig. 6.1, the threshold voltages of M1 and M2 (and similarly, M3 and M4) determine how well they turn on and hence determine the charge/discharge rate in read/write operations. This rate is closely

C. Gu (✉)
University of California, Berkeley, USA
e-mail: gcj@eecs.berkeley.edu

A. Singhee and R.A. Rutenbar (eds.), *Extreme Statistics in Nanoscale Memory Design*, Integrated Circuits and Systems, DOI 10.1007/978-1-4419-6606-3_6,

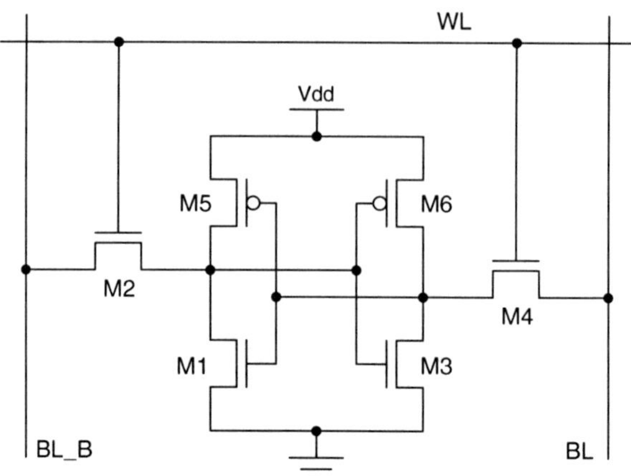

Fig. 6.1 SRAM 6T cell

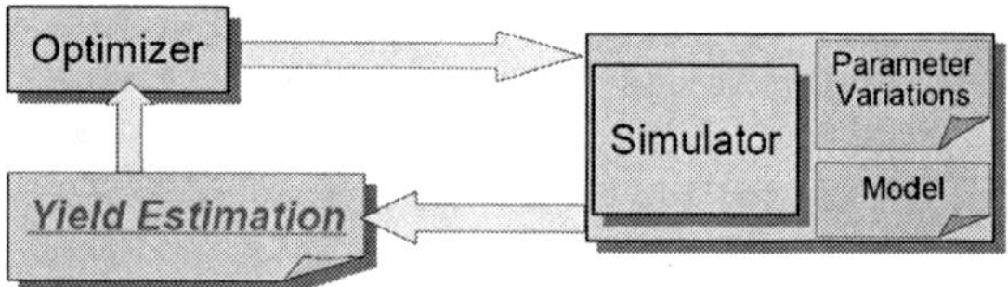

Fig. 6.2 Optimization loop in circuit design

related to SRAM read/write access time, and hence determines read/write frequency. Due to parameter variations, it is highly possible that some cells fail to finish the read/write operation within specified access time. As a result, the whole SRAM array circuit is defective, and has to be discarded. So it is crucial to predict what fraction of circuits is likely to fail because of variability, so that measures may be taken to increase the yield of correctly functioning circuits, to the extent possible via design and process changes. Reasonably accurate prediction of yield (e.g., within a percent or better) is important: over-estimation leads to fewer working components than expected, while under-estimation can lead to lowered performance as a result of design efforts to increase yield.

Due to its importance in the design process, the yield estimation problem is closely related to circuit optimization. Figure 6.2 shows a typical optimization loop for yield maximization in circuit design. In this loop, the simulator computes circuit performances of interest and their distributions, given device models and parameter distributions. Yield estimation is conducted around the simulator and the result is exported to an optimizer. The yield is one of the key quantities that guides the optimizer to achieve the "best" design point, which maximizes yield while optimizing performances.

In this chapter, we shall discuss numerical methods for yield estimation due to parameter variations. We first define the yield estimation problem in this section, and motivate the problem by SRAM circuit. In Sect. 6.2, we provide a survey of various approaches to the yield estimation problem, including statistical methods and deterministic methods. One of the deterministic methods, Yield Estimation via Nonlinear Surface Sampling (YENSS), turns out to be efficient in identifying the region (in the parameter space) that corresponds to circuits meeting all performance specifications/constraints, and in estimating the yield by computing the probabilistic hypervolume of that region. We discuss details of YENSS in Sect. 6.3. In Sect. 6.4, we show applications of YENSS on illustrative examples and an SRAM circuit.

6.1.1 Yield Estimation

We first define the yield estimation problem with a focus on its geometrical interpretations.

Briefly, the yield estimation problem is simply to compute the probability that the circuit will meet all specified performance constraints (bounds). Important inputs to a yield estimator include information about parameter variations (e.g., mean, variance, min/max bounds, and distributions), and performance specifications (e.g., min/max bounds). For example, a simple yield estimation problem with SRAM read access time as a performance metric can be stated as follows: given that the threshold voltages of MOSFETs in the SRAM cell satisfy uniform distributions on $[0. 1V, 0. 7V]$, calculate the probability that the SRAM read access time is less than 5 ns.

Given the complexities of the IC manufacturing process, performance specifications are typically much easier to obtain than information about parameter variations. In fact, characterizing parameter variations is a challenging problem in itself. Sometimes only min/max bounds, or mean values of the parameters, are available. If this happens, one typically makes assumptions about parameter distributions. For example, the uniform distribution and the (truncated) Gaussian distribution are commonly used [8].

From a geometrical view point, yield estimation is equivalent to a hypervolume computation problem [9] – this geometrical interpretation leads to a better understanding of the problem, and provides insights for numerical solution.

For purposes of illustration, suppose we have two independent parameters $\mathbf{p} = [p_1, p_2]$, with nominal values $p_{1\text{nom}}, p_{2\text{nom}}$, uniformly distributed over $[p_{1\text{min}}, p_{1\text{max}}]$ and $[p_{2\text{min}}, p_{2\text{max}}]$, respectively, as shown in Fig. 6.3. We assume that the circuit performs correctly at its nominal point, i.e., the circuit performance $f_{\mathbf{p}}$ is f_{nom} when the parameters are $(p_{1\text{nom}}, p_{2\text{nom}})$. As the parameters move away from the nominal design point, the circuit performance changes away from its nominal value. Therefore, in the parameter space, there exists a region D around the nominal design point where the performance remains acceptable. We call this region the *acceptable region* later on, as opposed to the *failure region*, i.e., the region in the min/max box excluding the acceptable region. The acceptable region is depicted by the interior of the closed curve in Fig. 6.3, with the portion within the min/max bounds

Fig. 6.3 Evaluating yield is equivalent to evaluating ratio of the area of the shaded part to the area of the box $[p_{1\min}, p_{1\max}]$ – $[p_{2\min}, p_{2\max}]$

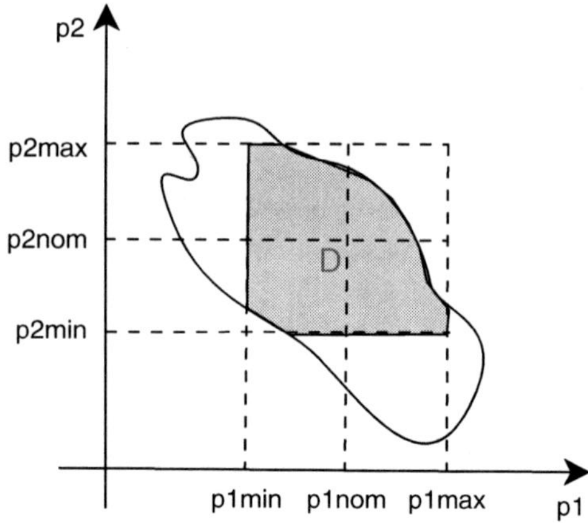

shown shaded. The yield is then the ratio of the area of region D to that of the box $[p_{1\min}, p_{1\max}] \times [p_{2\min}, p_{2\max}]$.

Mathematically, hypervolume computation is equivalent to calculating the integral

$$I = \int_D \frac{1}{(p_{1\max} - p_{1\min})(p_{2\max} - p_{2\min})} \, dp_1 dp_2. \tag{6.1}$$

Note that because p_1 and p_2 are uniformly distributed over $[p_{1\min}, p_{1\max}]$ and $[p_{2\min}, p_{2\max}]$, respectively, their probability density functions are $\frac{1}{p1\max - p1\min}$ and $\frac{1}{p2\max - p2\min}$, respectively. So the function to be integrated in (6.1) can be viewed as the joint probability density function of p_1 and p_2.

More generally, if the parameters $\mathbf{p}$ satisfy the joint probability density function $\pi(\mathbf{p})$, the yield, corresponding to the integral of the joint probability density function over the region D, is

$$I = \int_D \pi(\mathbf{p}) d\mathbf{p}. \tag{6.2}$$

We denote the integral (6.2) to be the *probabilistic hypervolume* (or probability mass) of region D.

If we think in terms of performance and parameter spaces, yield estimation is essentially a mapping from a region in performance space to a region in parameter space, the hypervolume of which constitutes the yield, as depicted by the lower arrow in Fig. 6.4. Hence, one way to solve the yield estimation problem is first to identify the boundary of this region in parameter space, and then use it to compute the probabilistic hypervolume. In this chapter, we will look into several methods for finding this boundary numerically.

Moreover, as these methods are dealing with boundaries in parameter and performance spaces, they can be applied to other similar problems involving

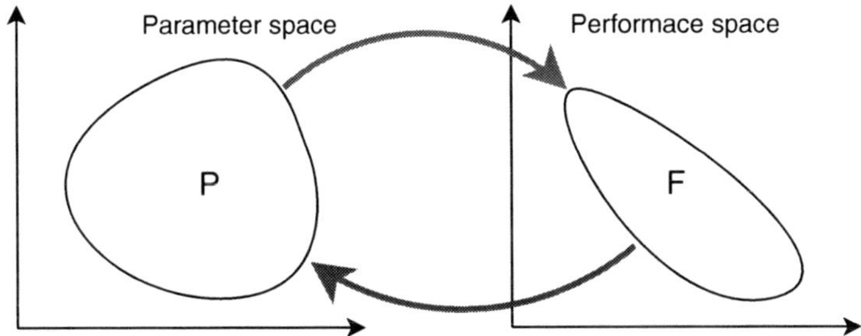

Fig. 6.4 Boundaries in parameter space and performance space

boundaries. For example, in the design centering (or yield maximization) problem [10–20], we are seeking a nominal design point[1] which maximizes the probability that circuits meet their specifications. Geometrically, we are looking for the point that maximizes the minimum distance to the boundary in parameter space, if independent and uniform parameter distributions are assumed. Being able to identify the boundary (or to compute points on the boundary) is an important subproblem for design centering.

As another example, the boundaries in performance space are used in the performance trade-off problem [21–23]. In this problem, we normally have several performance metrics to balance, say circuit speed and power consumption. For the parameters that are allowed to change, there exists a trade-off curve in the performance space – we can move on that trade-off curve to obtain the maximum performance of interest (e.g., maximize the speed within the power budget). Again, this requires that we find the boundary in the performance space, given a region in the parameter space – i.e., the upper arrow in Fig. 6.4. Similar to the design centering problem, the parameters used here are typically physical parameters.

To summarize, the yield estimation problem is equivalent to computing the probabilistic hypervolume of the region that correspond to working circuits. Technologies to identify the boundary of this region can serve as a useful first step in yield estimation, and can also be applied to other problems such as design centering and performance trade-off problems.

6.1.2   An Example: SRAM Read Access Failure

In SRAM cells, such as the 6T cell in Fig. 6.5a, several types of failures can occur due to variations of parameters such as the threshold voltages of the six MOSFETs [24]. For example: read access failure can be caused by an increase in cell read

---

[1]For the design centering problem, the parameters to be determined are normally physical parameters, such as transistor width W and length L; For the yield estimation problem, the parameters are "statistical" ones, such as threshold voltage V_t and channel length variation ΔL.

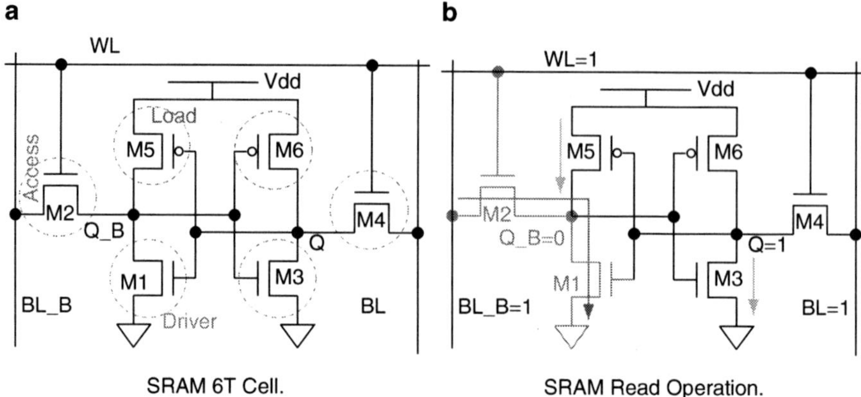

Fig. 6.5 SRAM 6T cell (**a**) and its read operation (**b**)

access time; data flipping during the read operation leads to an unstable read; write failure is the inability to write the data into a cell successfully within the write period; fluctuations of surrounding circuits/power supply can make the cell unable to hold data in standby mode, etc. Consider for example the yield loss caused by read access time failure in a 6T cell.

The SRAM 6T cell circuit is composed of a pair of cross-connected inverters (used for storing the data bit in the cell) and two pass transistors (used for accessing or writing the data from/to the SRAM cell). In Fig. 6.5a, $M2$ and $M4$ are the access transistors, and ($M1$, $M5$) and ($M6$, $M3$) form two cross-connected inverters. ($M1$, $M3$) and ($M5$, $M6$) are also commonly referred to as driver and load transistors, respectively. Q and Q_B are the nodes that actually store the data bit. BL and BL_B are the bit-lines which follow the cell data during the read operation, and provide the data to be written during the write operation. The access transistors are controlled by the wordline (WL), which enables/disables the SRAM cell to be read/ written.

To analyze the read operation of this SRAM cell, we assume, without loss of generality, that 1 is stored in the SRAM cell, i.e., $Q = V_{dd}$ and $Q\_B = 0$. During read operation, both bit-lines BL and BL_B are first precharged to V_{dd}. Then, at the assertion of wordline WL, bit-line BL_B starts to discharge through $M2$ and $M1$ (as shown in Fig. 6.5b via a path through M_2 and M_1), while bit-line BL stays at V_{dd}. Besides, subthreshold leakage paths through $M3$ and $M5$ are also present and affect the discharging current. Within the read access time (denote it as t_f), however, the resulting voltage difference between BL and BL_B (denote it as ΔBL) is small, and therefore a sense amplifier is used to amplify this signal to obtain a full swing $0 - V_{dd}$.

In order for the sense amplifier to be able to detect the bit-line voltage difference ΔBL (i.e., to differentiate signal 1 from signal 0), ΔBL has to reach a minimum value $\Delta BL_{\min}$. Therefore, a read access failure is said to occur if at the specified read access time t_f, the bit-differential ΔBL is less than the minimum required value $\Delta BL_{\min}$.

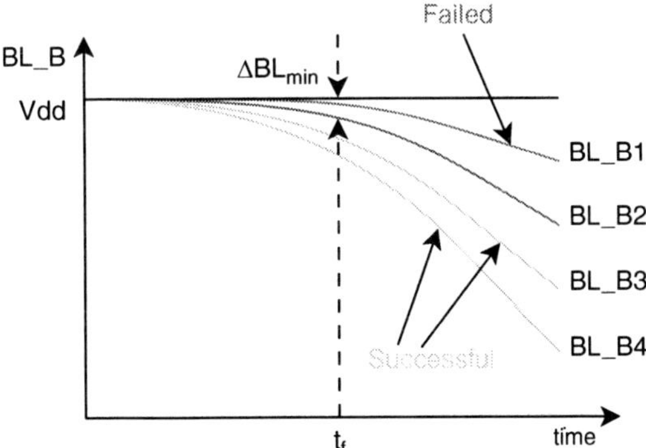

Fig. 6.6 ΔBL waveforms during read operation

Due to variations in the threshold voltages, the driving strengths of $M1$ and $M2$, as well as the leakage current through M_3 and M_5, change and affect the time required to generate the minimum voltage difference between BL and BL_B. For example, Fig. 6.6 shows several waveforms of BL_B in SRAM cells with different parameters, where t_f is the maximum read access time specified by the design. From Fig. 6.6, we see that if ΔBL is larger than the minimum bit-differential ΔBL_{min}, the read operation is successful (e.g., waveforms BL_B3 and BL_B4). Otherwise, the read operation is a failure (e.g., waveform BL_B2).

Therefore, the probability of SRAM failure due to read access time failure can be written as:

$$P_{\text{access–failure}} = P(\Delta BL(t_f) < \Delta BL_{min}), \qquad (6.3)$$

where $\Delta BL(t_f)$ is the bit-differential at time t_f. So the yield is computed by:

$$\text{Yield} = 1 - P_{\text{access–failure}} = P(\Delta BL(t_f) \geq \Delta BL_{min}). \qquad (6.4)$$

In the subsequent sections, we present several approaches to solve this yield estimation problem.

6.2   Approaches to Yield Estimation

Several approaches have been proposed for the yield estimation problem. They can be classified into two general categories: statistical methods and deterministic methods.

Statistical methods are mainly based on the law of large numbers and the central limit theorem []. They generate random samples in the parameter space, perform the simulation for each sample and count the number of samples that meet perfor mance specifications. In order to obtain a high accuracy of the results, they require a large number of samples in the parameter space, and consequently require a large number of simulations. However, as the number of samples needed for a certain accuracy does not increase dramatically when the number of parameters of interest increases, these methods can be well suited to problems which involve a very large number of parameters.

On the other hand, deterministic methods, or deterministic-statistical mixed methods, can generally be decomposed into two steps:

1. In the parameter space, identifying the acceptable region, i.e., finding the boundaries of this region;
2. Calculating the probability that parameters fall in the acceptable region.

In contrast to statistical methods, deterministic methods explicitly compute boundaries in the parameter space. Therefore, these methods are more attractive for various optimization procedures that need to utilize the boundary information.

Because of this, the accuracy of deterministic methods depends strongly on how well the boundaries are approximated. Usually piecewise linear approximation to the boundaries are used: identify a set of points on the boundary, and use linear interpolation to represent the boundary. The more points on the boundary, the more accurate yield estimation is. For problems that deal with a small number of parameters, and especially for problems that result in a nearly linear boundary in the parameter space, the boundary can usually be efficiently approximated by just a few points, which is much less than the number required for statistical methods. Deterministic methods are also well suited for such situations.

However, when there are large number of parameters, the computational cost of identifying boundaries can increase exponentially, due to the curse of dimensionality. In such cases, statistical methods are generally preferred; or deterministic methods are applied to obtain a coarse approximation of the boundary, which can be utilized to speed up statistical methods. Such methods can be considered statistical-deterministic mixed methods.

6.2.1 Statistical Methods

6.2.1.1 Monte-Carlo Methods

The simplest way to estimate yield is to apply the well known Monte-Carlo method [25–27]. Monte-Carlo methods are generally composed of the following four steps:

1. Sample parameters according to their statistical distributions (random number generation);
2. Simulate the circuit for each set of sampled parameters, and compute the circuit performances of interest;
3. Count the number of sampled circuits that meet all the performance specifications;
4. Calculate the ratio of this number to the total number of samples (calculating the yield).

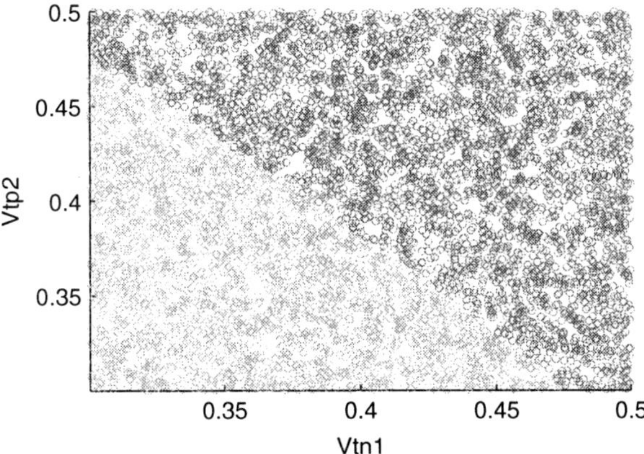

Fig. 6.7 Monte-Carlo simulation example. The *light samples* represent circuits that meet the specification, and the *dark samples* represent circuits that fail to work. The yield is calculated as the ratio of the number of *light samples* to the total number of samples

For example, Fig. 6.7 shows results from a Monte-Carlo simulation on a circuit, where the threshold voltages V_{th1}, V_{th2} of two MOSFETs are considered to be varying parameters. Random samples are first generated for V_{th1}, V_{th2}, and then simulations are performed for each (V_{th1}, V_{th2}) pair. The light samples represent circuits that meet the specification, and the dark samples represent circuits that fail to work. The yield is finally calculated as the ratio of the number of light samples to the total number of samples.

While Monte-Carlo methods have the advantages of simplicity and extremely general applicability, their main limitation is that they can require a very large number of samples as well as expensive simulations for those samples. According to the law of large numbers, if X is a random variable, μ is its mean, and σ^2 is its variance, we have

$$P(|\overline{X} - \mu| \geq \varepsilon) \leq \frac{\sigma^2}{N\varepsilon^2}, \tag{6.5}$$

where $\overline{X}$ is the sample average, N the number of samples, and ε represents how accurate the sample average approximates the real mean. Therefore, the smaller the ε is, the larger number of samples N needs to be taken.

Besides the large number of samples, the computation required for Monte-Carlo type yield estimation methods is significantly exacerbated if each individual simulation is expensive.

For example, SPICE-level transient simulations [28] are ideally run with tight tolerances (small time steps) and the best device models available, in order to extract accurate results such as timing information for digital circuits (e.g., delay of a critical path, setup/hold time of a latch/register, SRAM read/write time); RF analyses such as harmonic balance and shooting [29] are indispensable for circuits such as mixers and oscillators. These simulations are significantly more expensive than DC and AC analyses.

Whether such analyses are required depends on the circuit and performance of interest. While DC/AC analyses are usually enough to simulate linear circuits or circuits working in linear mode (such as operational amplifier circuits), accurate transient/RF analyses are essential for highly nonlinear circuits, such as SRAM cells and oscillators. For example, to find SRAM read/write times, transient simulations with very small time steps needs to be taken. To characterize the static noise margin (SNM) [7, 30] of an SRAM cell, the butterfly curve is typically identified, for which a series of DC analyses with respect to different DC inputs need to be performed.

Note that the types of simulations are not just confined to the applications just mentioned. For example, static timing analysis (STA) [31] and statistical static timing analysis (SSTA) [32] can be employed as a higher-level timing simulation instead of detailed transient analysis.

To summarize, the large number of samples that can be needed, and the expense of each simulation motivate alternatives to Monte-Carlo that can alleviate the computational cost of either or both aspects.

6.2.1.2   Improved Monte-Carlo Methods

In order to overcome the disadvantages of the classical Monte-Carlo method, various techniques have been proposed to improve its efficiency. These techniques basically try to sample data more intelligently so that the number of samples decreases without sacrificing accuracy.

The most widely used technique to reduce the number of samples is the class of variance reduction methods [26, 27]. For example, importance sampling [33, 34], stratified sampling [35, 36], Latin hypercube sampling[37], control variates [26, 27], antithetic variates [38], Rao-Blackwellization [26, 27] all fall in this category. The key idea of variance reduction methods is to reduce the variance of sample points – from (6.5), we know that for a fixed accuracy ε, the smaller the variance σ^2, the smaller number of samples are needed. Some of these techniques are detailed in other chapters.

6.2.2   Deterministic/Mixed Methods

As mentioned before, denote the acceptable region by D, the probability density functions of parameters $\mathbf{p}$ by $\pi(\mathbf{p})$; then, the yield estimation problem is to calculate the integral

$$I = \int_D \pi(\mathbf{p})\mathrm{d}\mathbf{p}. \tag{6.6}$$

The Monte-Carlo method can be viewed as a way to approximate this multi-dimensional integral. On the other hand, if we can identify region D, i.e., find the boundaries of the region D, numerical integration methods can be employed to compute (6.6) directly.

The boundaries of D can be represented as the solution of a nonlinear equation. Suppose there are N_p parameters $\mathbf{p}$, i.e., the parameter space is $\mathbb{R}^{N_p}$, then D is a N_p-dimensional closed region in $\mathbb{R}^{N_p}$, and its boundary is a $(N_p - 1)$-dimensional manifold. Also suppose there are N_f performances of interest $\mathbf{f_p}$, which depend on parameters $\mathbf{p}$. Then, the boundary for each performance constraint is defined by a *scalar* nonlinear equation[2]

$$h(\mathbf{p}, \mathbf{f_p}) = 0 \tag{6.7}$$

which states the relationship between $\mathbf{p}$ and $\mathbf{f_p}$.

We now review several methods that solve for or approximate this boundary, prior to using the boundary to perform yield estimation.

6.2.2.1 Yield Estimation by Simplicial Approximation

In Sect. 6.2.1, we have seen that Monte-Carlo methods with variance reduction techniques reduce the computational cost only by reducing the number of samples. By utilizing simplicial approximations of the boundary in parameter space, one can further speed up the procedure by also reducing the computational cost of simulating each individual sample. Later, we outline the basic structure of this algorithm[3] [39, 40].

1. Define the maximum number of samples of Monte-Carlo method to be $N_{\max}$, and the number of working circuits to be N_{good};
2. Identify a few points on the boundary, and build up the linear constraints to represent a crude approximation to the boundary;
3. Start Monte-Carlo sampling:
   (a) If the sample lies inside the boundary (satisfies all linear constraints), $N_{\text{good}} = N_{\text{good}} + 1$;
   (b) Else, perform a full simulation to check if the sample lies inside the boundary, and apply line search to solve for another point on the boundary. If it passes the check, $N_{\text{good}} = N_{\text{good}} + 1$;
   (c) Update the boundary (by adding/deleting linear constraints that define the boundary);
4. Compute the yield: Yield $= \dfrac{N_{\text{good}}}{N_{\max}}$.

---

[2]This is the case when the boundary is defined by one performance constraint. The multiple-constraint case will be discussed in Sect. 6.3.1.

[3]There are a few variations of this algorithm, but the basic structures are almost the same.

In [39, 40], the boundary in parameter space is approximated by a series of linear constraints, i.e., a simplicial approximation. The utility of the approximate linear constraints to the boundary is that testing whether a point is in the acceptable region is sped up significantly for most (but not all) Monte-Carlo samples – in order to judge whether a sample lies inside the boundary or not, one does not need to simulate the sample, but just test the linear constraints.

For those that fail this fast check, a more computationally expensive check, involving a linear search between the approximate linear constraints and the real boundary, is performed. As a by-product of this check, a new sample on the real boundary is obtained, and therefore the boundary is refined by adding/deleting linear constraints.

As we normally expect a high yield, most samples fall inside the boundary; therefore, with only a very coarse approximation of the boundary, we save the computational cost of simulating most samples. The price paid is the computation needed to identify the boundary. To identify a new point on the boundary, line search method usually takes 3–4 iterations to converge, which correspond to 3–4 simulations of the circuit.

One problem with this approach is that the number of linear constraints, or the number of points on the boundaries, increases with the number of parameters to be considered if the same accuracy of the boundary approximation is desired. More-over, the points on the boundary are generated randomly – if a sample point fails to meet the linear constraints, a new point, on the boundary and close to this sample point, is searched for by the algorithm. That is to say, it is not guaranteed that the points that approximate the boundary are well-spaced, or that the linear constraints will approximate the acceptable region well – the shape of the approximated boundary can be quite different from that of the real boundary if these points are clustered together in a small region.

There is also an important assumption behind this approach – the boundary separating the acceptable/failure regions must be convex. It is this assumption that enables us to perform a fast check on a new sample, instead of performing a full simulation. However, this assumption is not always true in reality. Indeed, the boundary for SRAM read access failure, as we show in Fig. 6.20b, is not convex. When this assumption is not met, the algorithm might incur error in the result.

While simplicial approximation essentially uses a set of linear constraints to approximate the boundary, there is no limit for choosing the form of constraint equations. Indeed, similar to simplicial approximation, there are methods that use quadratic approximations [41] to the boundary. However, third-order or higher-order approximations are typically not employed because they are computationally much more expensive.

6.2.2.2 Yield Estimation by Worst-Case Distance Approximation

To alleviate the problem of dramatic increase of number of points with the number of parameters in simplicial approximation methods, a simpler method based on worst-case distance approximation [42] has been proposed. The parameters that have the *worst-case distance* are defined to be those that have the smallest metric distance from the nominal parameter point. These points have the highest probability of causing performance constraint to be violated. Based on this key idea, one worst-case parameter point with respect to each performance constraint is searched for by applying numerical optimization methods. Therefore, the boundary is approximated with many fewer points.

The basic flow of the algorithm is:

1. For each performance constraint, define a corresponding optimization problem that finds the worst-case parameters;
2. Solve each optimization problem, and obtain a point for each optimization problem on the boundary;
3. Use the tangent hyperplane at each point to approximate the boundary;
4. Calculate a yield estimate for each performance constraint using an analytical formula; or, calculate a yield estimate considering all performance constraints simultaneously by the Monte-Carlo method.

There are several options for the distance d between two points P_1 and P_2 in the parameter space. The most common definition is the Euclidean distance between P_1 and P_2. Equivalently, it is the L_2-norm of the vector difference between two points

$$d^2(\mathbf{p}_1, \mathbf{p}_2) = \|\mathbf{p}_1 - \mathbf{p}_2\|_2^2 = (\mathbf{p}_1 - \mathbf{p}_2)^T (\mathbf{p}_1 - \mathbf{p}_2), \tag{6.8}$$

where $\mathbf{p}_1$ and $\mathbf{p}_2$ are the coordinates of P_1 and P_2, respectively.

As the norms are equivalent [43], other norms such as the weighted L_2-norm

$$d^2(\mathbf{p}_1, \mathbf{p}_2) = (\mathbf{p}_1 - \mathbf{p}_2)^T C (\mathbf{p}_1 - \mathbf{p}_2), \tag{6.9}$$

the L_∞-norm

$$d(\mathbf{p}_1, \mathbf{p}_2) = \max |\mathbf{p}_1 - \mathbf{p}_2| \quad \text{(element-wise maximum)} \tag{6.10}$$

etc., can be used.

They also have intuitive geometrical interpretations, as shown in Fig. 6.8. For example, in the parameter space, the L_2-norm worst-case distance correspond to a hyper sphere, the weighted L_2-norm corresponds to a hyper-ellipsoid, and the L_∞-norm corresponds to a hyper-rectangle (orthotope).

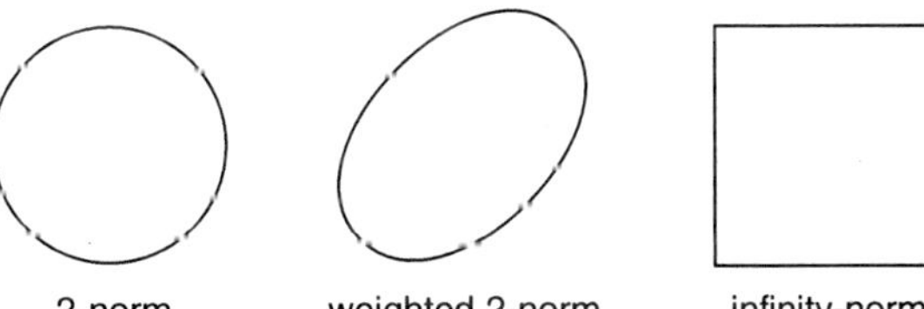

Fig. 6.8 Geometrical interpretations of norms

After the worst-case point on the boundary is returned from the nonlinear optimizer, the tangent hyperplane on this point is calculated by performing a sensitivity analysis. This tangent hyperplane separates the parameter space into two parts, one of which is assumed to be the failure region, and the other to be the acceptable region. This makes rough yield estimation an easy problem, because the multi-dimensional integration degrades to one-dimensional integration, whose computational cost is almost negligible.

When multiple constraints are considered simultaneously, the Monte-Carlo method is again employed to compute the integration, just as in simplicial approximation methods.

This method is good in situations where only rough yield estimation is desired. Because for each performance constraint, there can be infinite number of parameters that achieve the worst-case performance, choosing one point for each constraint and using the tangent hyperplane to approximate the boundary is just a coarse approximation.

6.2.2.3   Yield Estimation by Ellipsoidal Approximation

Ellipsoidal approximation methods [11, 44] are a trade-off between simplicial approximation methods and worst-case distance approximations: they are computationally cheaper than simplicial approximation methods, and are more accurate than worst-case distance approximations.

Unlike the previous two methods, which use local approximations to the boundary, ellipsoid approximation methods iteratively identify a hyper-ellipsoid to obtain a global approximation to the boundary. Variants of ellipsoid approximation differ in the way the hyper-ellipsoid is defined and calculated. For example, in [11], the ellipsoid is defined to be the largest possible one whose axial end points do not violate performance constraints. As another example, the method of [44] generates a sequence of hyper-ellipsoids of decreasing hyper-volume until a final ellipsoid satisfying a specified hyper-volume reduction ratio is obtained.

There are several advantages to using ellipsoidal approximations: firstly, the representation of a hyper-ellipsoid in a high-dimensional space is simple, i.e., it is a quadratic equation

$$(\mathbf{p} - \mathbf{p}_0)^T C^{-1} (\mathbf{p} - \mathbf{p}_0) = 1, \tag{6.11}$$

where $\mathbf{p}_0$ is the center of the ellipsoid and C is a positive definite matrix. This avoids the storage limit that is a bottleneck for simplicial approximation methods.

Secondly, the hyper-volume of the hyper-ellipsoid has an elegant formula

$$V = V_0 \sqrt{|C|}, \tag{6.12}$$

where V_0 is the hyper-volume of the unit sphere. This makes it possible to use deterministic methods, rather than the Monte-Carlo method, to perform the numerical integration in the last step.

6.2.2.4 Yield Estimation by Euler–Newton Curve Tracing

Euler–Newton curve tracing [45] is another way to find the boundaries of the acceptable region D, and is best suited to the two-parameter case. Euler–Newton curve tracing is a kind of continuation method [46] to trace a curve in the parameter space. The idea is to start from a point on the boundary curve, and then to follow the curve by using the Newton–Raphson method to find adjacent points. The outline of the algorithm is as follows:

1. Use the Moore–Penrose pseudo-inverse Newton–Raphson (MPNR) method (discussed in Sect. 6.3.5) to solve for one point on the boundary;
2. Euler–Newton curve tracing:
   (a) Perform an Euler step (see later) to predict an approximate adjacent point on the boundary;
   (b) Using the Euler prediction as the initial guess for MPNR, solve for the next point on the boundary;
   (c) Repeat this prediction–correction scheme until the boundary is identified;
3. Calculate the yield by performing numerical integration of the region inside the boundary.

In this algorithm, the MPNR method is an iterative method to solve underdetermined nonlinear algebraic equations. It starts from an initial guess to the solution, and iteratively refines the solution until a required accuracy is reached. The Euler step is a prediction for a new point on the boundary. It follows the tangent at a point on the curve, and feeds the result into MPNR as the initial guess. Therefore, the Euler–Newton curve tracing method is basically a prediction–correction method [43], where the Euler step serves as the predictor and the MPNR method serves as the corrector. In the last step, there are various choices for the numerical integration method [43], such as trapezoidal rule, Simpson's rule, etc.

As an example, we consider the SRAM read access failure problem regarding variations of threshold voltages of $M1$ and $M2$. Figure 6.9a shows the parameter space of V_{th1} and V_{th2}. Suppose the curve in Fig. 6.9a is the boundary separating the acceptable region and the failure region. In the first step, we solve (6.7) to obtain one point on the boundary using MPNR – given an initial guess at point A, MPNR converges to point B on the boundary. The initial guess can be given heuristically, such as an approximate solution from designer's intuition. In MPNR, the computation of (6.7) is through a full-SPICE transient simulation, and its Jacobian matrix can be obtained by performing a sensitivity analysis (discussed in Sect. 6.3.6) or otherwise approximated.

After the starting point is obtained, we take an Euler step (denoted by the arrow $A_N \rightarrow B_E$ in Fig. 6.9b) to predict the next point on the curve. For example, starting from A_N, the Euler step predicts the next point to be B_E. The Euler prediction usually gives a very good initial guess for the next point on the curve, and therefore it makes Newton–Raphson achieve its local quadratic convergence. This prediction–correction procedure is repeated until the boundary is identified. In practice, this prediction–correction procedure turns out to be quite efficient – only

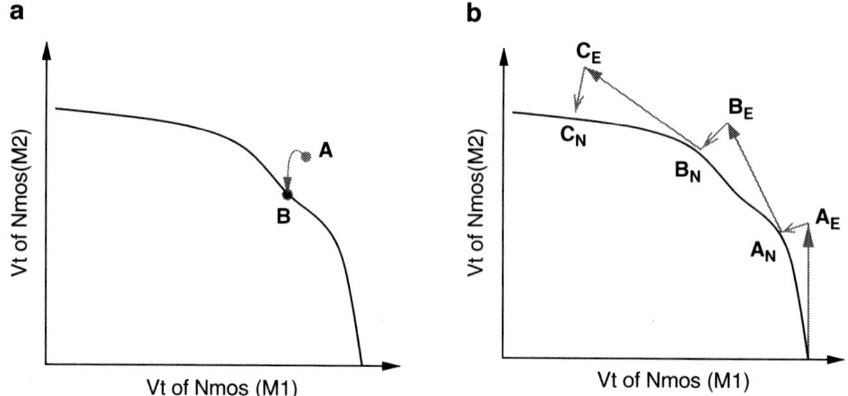

Fig. 6.9 (**a**) Convergence of a point via Newton–Raphson on the boundary. (**b**) Illustration of Euler–Newton curve tracing

up to 2–3 Newton iterations are needed to identify a new point on the boundary. Finally, a numerical integration formula, such as the trapezoidal rule [43], is employed to calculate the (probabilistically weighted) area inside the boundary.

However, the main drawback of the Euler–Newton curve tracing method is also obvious – it cannot be easily generalized to high-dimensional problems. For example, when three parameters are taken into consideration, the Euler step must be defined in two directions.

6.2.2.5 YENSS Sampling

Another method, YENSS will be discussed in detail in the next section. Like the Euler–Newton method, it also fully characterizes the boundary in the parameter space, and calculates the probabilistic hypervolume of the region inside the boundary. For the underlying computations involved in finding points on the boundary, YENSS supports two versions of the algorithm, based on Moore–Penrose Newton–Raphson and line search [47], respectively – the latter can be useful in situations where parameter derivative/sensitivity information, required by Newton–Raphson, is difficult or expensive to obtain.

Importantly, unlike the Euler–Newton curve tracing method, YENSS generalizes to high dimensions. However, if there are many parameters to be considered, fully characterizing a high-dimensional boundary manifold is also very computationally expensive. To alleviate this problem, YENSS uses an adaptive technique to calculate only as many points on the boundary as needed to provide an estimate of yield to a desired accuracy. By doing so, YENSS avoids redundant/unnecessary simulations in identifying the boundary manifold. This is especially useful when many parameters are considered simultaneously: the accuracy is controlled by the "yield volume" increment at each iteration, compared to $\frac{1}{\sqrt{N}}$ accuracy improvement in Monte-Carlo simulation, where N is the number of samples.

A crucial feature of YENSS, which is central to its efficiency for problems with many parameters, is its automatic exploitation of *linearity* of the boundary. If the boundary is perfectly linear (e.g., a plane embedded in 3-dimensional parameter space), the computational expense of YENSS increases only linearly with the number of parameters N_p. Because it adaptively estimates the probabilistic volume, the computation needed by YENSS increases gracefully to handle boundaries that are not linear. As failure/acceptable boundaries in parameter space tend to be close to linear for many practical problems, YENSS can achieve great speedups over Monte-Carlo.

6.2.2.6 Normal Boundary Intersection: The Reverse Problem

The deterministic methods we have discussed so far, including simplicial approximation, worst-case distance approximation, ellipsoidal approximation, Euler–Newton curve tracing and YENSS, are all concerned with boundaries in the parameter space. Given a region of desired (e.g., worst-case) performances of the circuit, the corresponding boundary in the parameter space is determined, as depicted by the lower arrow in Fig. 6.4. Techniques that solve the reverse problem (i.e., given a region in the parameter space such as a hypercube, find the corresponding boundary in the performance space, as depicted by the upper arrow in Fig. 6.4), also exist, and are important in performance optimization and trade-off analysis.

While this is a different class of problem from the yield estimation problem considered here, we note that the method of [21] uses a boundary approximation technique in the performance space that is similar to the methods used to solve for the boundary in YENSS and the simplicial approximation. It uses a search scheme (termed *Normal Boundary Intersection* (NBI)) to find equally spaced points on the boundary in the performance space. Instead of using a root-finding algorithm to identify the boundary, NBI formulates the problem as a sequential quadratic programming problem [47], and uses numerical optimization methods to solve for the boundary. As this method is also dealing with the boundaries, it can also be potentially adapted to solve the yield estimation problem.

Several core underlying subroutines are similar in these methods. For example, simplicial, worst-case distance, and ellipsoidal approximation methods all require a last step of Monte-Carlo simulation; simplicial approximation and nonlinear surface sampling methods both involve a line search step to identify a new point on the boundary, etc.

In the rest of the chapter, we will focus on YENSS in greater detail, in the context of which we will point out several reusable subroutines (such as principal component analysis, Newton–Raphson, line search, transient sensitivity analysis) that are used in other yield estimation methods.

6.3 Computing the Boundary and Probabilistic Hypervolumes

6.3.1 Basic Idea and Geometrical Explanation

The basic procedure of YENSS is the same as that of other deterministic yield estimation methods, and consists of two steps:

1. Identify the boundary in the parameter space (approximate the boundary manifold by hyper-triangular patches);
2. Compute the probabilistic hypervolume of the region inside the boundary manifold.

To illustrate the idea, we first consider the following simple problem, after which we deal with generalizations to more complex situations. Referring again to Fig. 6.3, suppose we have two independent parameters p_1 and p_2, with nominal values $p_{1\text{nom}}$, $p_{2\text{nom}}$, uniformly distributed over $[p_{1\text{min}}, p_{1\text{max}}]$ and $[p_{2\text{min}}, p_{2\text{max}}]$, respectively. As a result, in YENSS, the boundary curve, which is defined by a scalar nonlinear equation $h(\mathbf{p}, \mathbf{f_p}) = 0$, is first identified, and then the yield is computed to be the ratio of the area of region D to the area of the min/max box

$$\text{Yield} = \frac{\text{Area}_D}{(p_{1\text{max}} - p_{1\text{min}})(p_{2\text{max}} - p_{2\text{min}})}. \tag{6.13}$$

6.3.1.1   Extension to Multiple Parameters

The earlier notion is easily generalized to the case of multiple parameters. For the case of three parameters, the boundary is a 2-dimensional surface in the 3-dimensional parameter space, and the yield is the normalized volume of the 3-dimensional region that corresponds to acceptable performance in parameter space. For example, Fig. 6.10 shows a 3-dimensional parameter space, in which the boundary manifold is approximated by seven samples on it. The yield then is

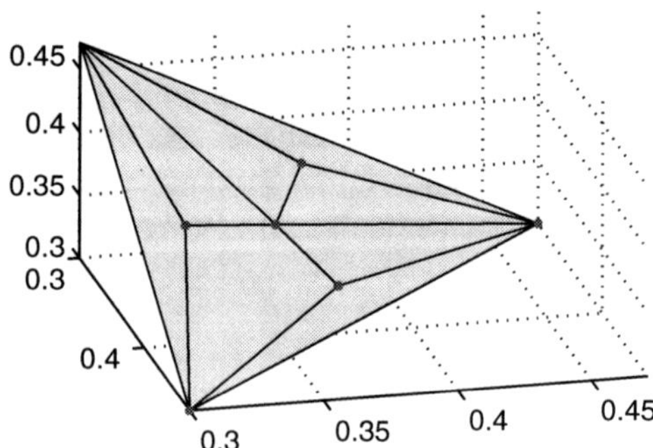

Fig. 6.10 Illustration of YENSS for three parameters

computed to be the ratio of the volume under this surface to the volume of the min/max box.

For the case of N_p parameters, the boundary is a $(N_p - 1)$-dimensional manifold that is defined by a scalar equation in the N_p-dimensional parameter space; the yield is the normalized hypervolume of the corresponding N_p-dimensional acceptable region in the parameter space.

6.3.1.2 Extension to Multiple Constraints

The earlier problem is an example where there is only one performance constraint. In this case, the boundary manifold is defined by one scalar equation:

$$h(\mathbf{p}, \mathbf{f_p}) = 0, \tag{6.14}$$

and the acceptable region (the region inside the boundary) is correspondingly defined by one scalar inequality

$$h(\mathbf{p}, \mathbf{f_p}) \leq 0. \tag{6.15}$$

A straightforward generalization to the case of multiple constraints is as follows: if there are N_c constraints, the region inside the boundary is defined by N_c inequalities

$$h_1(\mathbf{p}, \mathbf{f_p}) \leq 0,$$
$$h_2(\mathbf{p}, \mathbf{f_p}) \leq 0,$$
$$\cdots,$$
$$h_{N_c}(\mathbf{p}, \mathbf{f_p}) \leq 0, \tag{6.16}$$

and the boundary is defined to be the common interior of the manifolds defined by $h_i(\mathbf{p}, \mathbf{f_p}) = 0, \forall i \in [1, N_c]$.

For example, Fig. 6.11 shows an example with three constraints in the 2-dimensional parameter space. Three stand-alone boundaries correspond to the boundaries defined by $h_1(\mathbf{p}, \mathbf{f_p}) = 0$, $h_2(\mathbf{p}, \mathbf{f_p}) = 0$, and $h_3(\mathbf{p}, \mathbf{f_p}) = 0$, respectively. Then the final boundary is the common interior of all the three separate boundaries, and the acceptable region is the shaded region D. The computational procedure to identify this compound boundary manifold will be discussed in Sect. 6.3.4.

6.3.1.3 Extension to Handling Probability Distributions

In practice, the parameters are not likely to all have uniform distributions. Other probability distributions, such as Gaussian and truncated Gaussian distributions

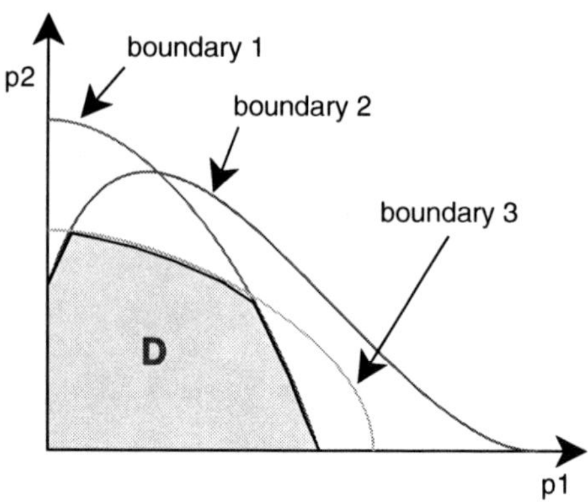

Fig. 6.11 Illustration of multiple constraints

[48, 49], are often used to model parameter variability. In this case, the hyper-volume inside the boundary in the parameter space no longer represents the yield.

There are two ways in which one can still calculate yield based on the boundary. The first way is obvious from the formula Yield $= \int_D \pi(\mathbf{p})\mathrm{d}\mathbf{p}$. That is, using numerical integration of the probability density function $\pi(\mathbf{p})$ to compute the probabilistic hypervolume (or probability mass) of region D.

The second way is to transform the parameter axes $\mathbf{p}$ so that the transformed parameters $\hat{\mathbf{p}}$ are uniformly distributed. In order to do this, recall how to generate a random variable Y of probability density function $f_Y(y)$ from a random variable X which is uniformly distributed on [0, 1].

Suppose y is a monotonically increasing function of x, $y = g(x)$, i.e., we have an inverse function $x = g^{-1}(y)$.

By definition, we have

$$P(x < X < x + \mathrm{d}x) = f_X(x)\mathrm{d}x, \tag{6.17}$$

$$P(y < Y < y + \mathrm{d}y) = f_Y(y)\mathrm{d}y. \tag{6.18}$$

On the other hand, because $y = g(x)$, when x varies from x to $x + \mathrm{d}x$, y varies from $y = g(x)$ to $y + \mathrm{d}y = g(x) + g'(x)\mathrm{d}x$ (see Fig. 6.12). Therefore, the probability that $X \in [x, x + \mathrm{d}x]$ should equal the probability that $Y \in [g(x), g(x) + g'(x)\mathrm{d}x]$. So, if $\mathrm{d}y = g'(x)\mathrm{d}x$, we have

$$P(x < X < x + \mathrm{d}x) = P(y < Y < y + \mathrm{d}y). \tag{6.19}$$

Inserting (6.17) and (6.18) into (6.19), we obtain

$$f_X(x)\mathrm{d}x = f_Y(y)\mathrm{d}y. \tag{6.20}$$

Because X is uniformly distributed on [0, 1], $f_X(x) = 1$ on [0, 1]. Therefore,

$$x = \int_{-\infty}^{y} f_Y(y)\mathrm{d}y = F_Y(y), \tag{6.21}$$

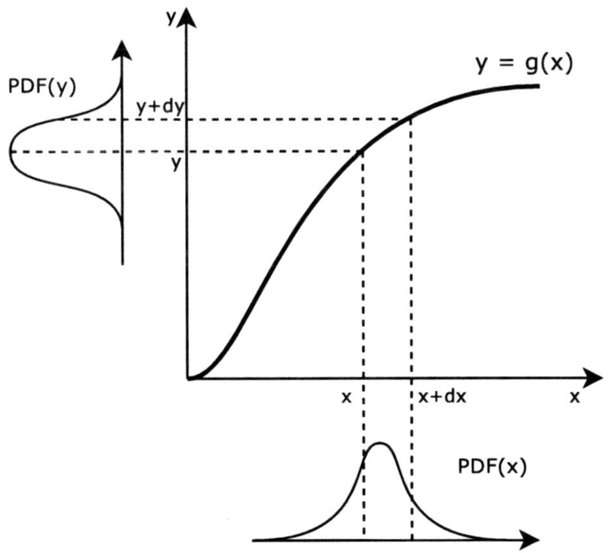

Fig. 6.12 Transformation of probability density function

or

$$y = F_Y^{-1}(x), \tag{6.22}$$

where $F_Y(y)$ is the cumulative density function of Y. Therefore,

$$g(x) = F_Y^{-1}(x). \tag{6.23}$$

As the cumulative density function $F_Y(y)$ is a monotonically increasing function, its inverse is also monotonically increasing, and our assumption regarding the monotonicity of $g(x)$ is validated.

Hence, if we have a random variable x uniformly distributed on $[0, 1]$ then $F_Y^{-1}(x)$ is a random variable with PDF $f_Y(y)$ [2].

Similarly, here we have an inverse problem to solve, i.e., we have a random variable Y with PDF $f_Y(y)$, and we want to compute the corresponding random variable that is uniformly distributed on $[0, 1]$. From the previous derivation,

$$x = F_Y(y) \tag{6.24}$$

is the random variable we want.

So, given parameters $\mathbf{p} = [p_1, \ldots, p_{Np}]$, we first transform $\mathbf{p}$ to their cumulative density functions $\hat{p}_i = F_i(p_i)$, $i \in [1, N_p]$, such that the scaled parameters $\hat{\mathbf{p}} = [\hat{p}_1, \ldots, \hat{p}_{N_p}]$ are uniformly distributed. For example, as shown in Fig. 6.13, the axes p_1 and p_2 in Fig. 6.3 are replaced by $\hat{p}_1 \equiv F_1(p_1)$ and $\hat{p}_2 \equiv F_2(p_2)$,

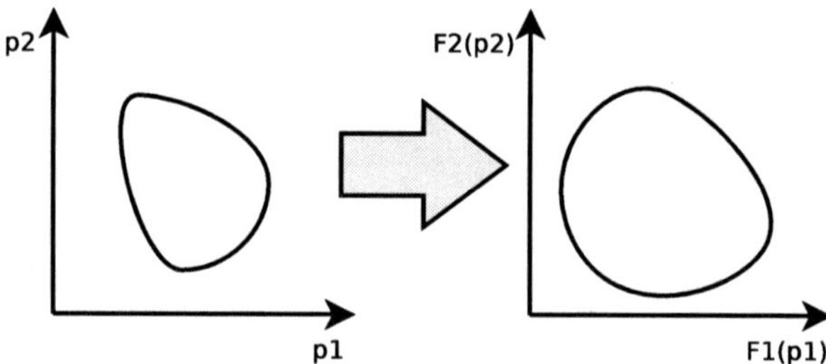

Fig. 6.13 Handling parameter probability distributions

respectively, where F_1 and F_2 are the cumulative distribution functions of the two parameters. Accordingly, the boundary in the original parameter space $\mathbf{p} \in \mathbb{R}^{N_p}$ is effectively mapped to a different boundary in the probabilistically deformed parameter space $\hat{\mathbf{p}} \in \mathbb{R}^{N_p}$. As $\hat{\mathbf{p}}$ are uniformly distributed, and if we assume that parameters $\mathbf{p}$ ($\hat{\mathbf{p}}$) are independent then the hypervolume of the region inside this new boundary is the yield.

The second method is preferred in YENSS because YENSS computes the yield by computing hypervolume of the acceptable region, which relies on the assumption of uniform distributions of parameters.

6.3.1.4   Extension to Handling Correlations of Parameters

The assumption that parameters are independent (and hence uncorrelated) is also important for the hypervolume calculated to be the yield. But this assumption is too strong to be true in reality. For example, W and L of a transistor are typically closely correlated with each other. So in the preprocessing step, we also need to de-correlate the parameters, and this can be done by principal component analysis (PCA) [50].

To see how PCA de-correlates parameters, we shall start with the definition of correlation between parameters/random variables. Suppose there are N_p parameters $\mathbf{p} = [p_1, \ldots, p_{N_p}]^T$, and $E[\mathbf{p}] = [\mu_1, \ldots, \mu_{N_p}]^T$. Here $E[\mathbf{p}]$ can be viewed as the nominal design parameters, and $\Delta\mathbf{p} = \mathbf{p} - E[\mathbf{p}]$ are zero-mean random variables that account for random parameter variations.

The covariance matrix R for $\Delta\mathbf{p}$ is defined by:

$$R = E[\Delta\mathbf{p}\Delta\mathbf{p}^T] = \begin{bmatrix} E[\Delta p_1 \Delta p_1] & E[\Delta p_1 \Delta p_2] & \cdots & E[\Delta p_1 \Delta p_{N_p}] \\ E[\Delta p_2 \Delta p_1] & E[\Delta p_2 \Delta p_2] & \cdots & E[\Delta p_2 \Delta p_{N_p}] \\ \vdots & \ddots & \ddots & \vdots \\ E[\Delta p_{N_p} \Delta p_1] & E[\Delta p_{N_p} \Delta p_2] & \cdots & E[\Delta p_{N_p} \Delta p_{N_p}] \end{bmatrix}. \tag{6.25}$$

Then, by definition, parameters $\mathbf{p}$ are uncorrelated if and only if R is diagonal.

In PCA, we try to find a linear transformation $\Delta\mathbf{q} = V\Delta\mathbf{p}$, so that the covariance matrix $R_{\mathbf{q}}$ for $\Delta\mathbf{q}$ is diagonal. As a result, we compute

$$\mathbf{q} = V\mathbf{p} = V(E(\mathbf{p}) + \Delta\mathbf{p}) = E(V\mathbf{p}) + \Delta\mathbf{q}. \tag{6.26}$$

As $\Delta\mathbf{q} = V\Delta\mathbf{p}$ are uncorrelated, and $E(V\mathbf{p})$ is constant, $\mathbf{q}$ are uncorrelated. Inserting $\Delta\mathbf{q} = V\Delta\mathbf{p}$ into the definition of $R_{\mathbf{q}}$, we obtain

$$R_{\mathbf{q}} = E[\Delta\mathbf{q}\Delta\mathbf{q}^T] = E[V\Delta\mathbf{p}\Delta\mathbf{p}^T V^T] = VE[\Delta\mathbf{p}\Delta\mathbf{p}^T]V^T = VRV^T. \tag{6.27}$$

The key step in achieving de-correlation is to eigen-decompose [51] the matrix R

$$R = P\Lambda P^{-1} = P\Lambda P^T, \tag{6.28}$$

where $P^{-1} = P^T$ because R is symmetric.

Inserting (6.28) into (6.27), we have

$$R_q = VP\Lambda P^T V^T. \tag{6.29}$$

Choosing $V = P^T$ then makes R_q a diagonal matrix. Therefore, the transformed parameters $\mathbf{q} = P^T\mathbf{p}$ are uncorrelated.

PCA successfully provides a means to remove second-order correlation between parameters. However, independence of parameters are still not guaranteed after PCA. If no higher-order correlations are present, or if higher-order correlations are negligible, the independence of parameters remains a reasonable assumption. This assumption is typically made in practice.

As an example of PCA, Fig. 6.14 shows 1,000 samples of parameters $\mathbf{p} = [p_1, p_2]^T$ which are generated from two independent Gaussian variables $\mathbf{q} = [q_1, q_2]^T$. Here we choose $q_1 \sim N(0, 1^2)$ and $q_2 \sim N(0, 3^2)$, and $p_1 = \frac{2}{\sqrt{5}}q_1 - \frac{1}{\sqrt{5}}q_2$, $p_2 = \frac{1}{\sqrt{5}}q_1 + \frac{2}{\sqrt{5}}q_2$. Therefore, we have

$$\begin{bmatrix} q_1 \\ q_2 \end{bmatrix} = \begin{bmatrix} \dfrac{2}{\sqrt{5}}p_1 + \dfrac{1}{\sqrt{5}}p_2 \\ -\dfrac{1}{\sqrt{5}}p_1 + \dfrac{2}{\sqrt{5}}p_2 \end{bmatrix}. \tag{6.30}$$

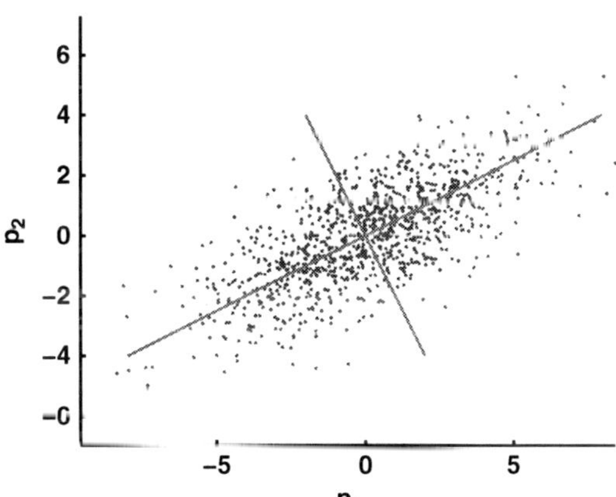

Fig. 6.14 Illustration of PCA

We assume only the covariance matrix for **p**, which can be approximated from the samples, is available, and perform PCA to rediscover the uncorrelated random variables q_1 and q_2.

Starting with the covariance matrix for **p**

$$
R = \begin{bmatrix} \dfrac{13}{5} & -\dfrac{16}{5} \\ -\dfrac{16}{5} & \dfrac{37}{5} \end{bmatrix},
\tag{6.31}
$$

we perform eigen-decomposition of R, and obtain $R = P \Lambda P^T$, where

$$
P = \begin{bmatrix} \dfrac{2}{\sqrt{5}} & -\dfrac{1}{\sqrt{5}} \\ \dfrac{1}{\sqrt{5}} & \dfrac{2}{\sqrt{5}} \end{bmatrix}.
\tag{6.32}
$$

Therefore, the uncorrelated variables are

$$
V\mathbf{p} = P^T\mathbf{p} = \begin{bmatrix} \dfrac{2}{\sqrt{5}}p_1 + \dfrac{1}{\sqrt{5}}p_2 \\ -\dfrac{1}{\sqrt{5}}p_1 + \dfrac{2}{\sqrt{5}}p_2 \end{bmatrix}
\tag{6.33}
$$

which coincide with q_1 and q_2. This example shows that PCA is able to automatically identify uncorrelated random variables.

Up to this point, we have shown that the idea of computing yield by probabilistic hypervolumes can be generalized to multiple parameters and multiple constraints, and that nonuniformly distributed and/or correlated parameters can, in typical situations, be mapped into uniformly distributed, uncorrelated parameters. Therefore, in the following sections, we only focus on the core problem which involves only uniformly distributed, uncorrelated parameters.

6.3.2 YENSS Algorithm Outline

The overall algorithm flow of YENSS is outlined later, and we explain and develop each step in detail in the following sections.

1. Preprocessings:
   (a) Perform a principal component analysis to de-correlate parameters (Sect. 6.3.1);
   (b) Transform parameters so that they are all uniformly distributed on [0, 1] (Sect. 6.3.1).

2. Formulate the boundary manifold as a nonlinear scalar equation (performance constraint equation) in the parameter space:

$$h(\mathbf{p}; \mathbf{f_p}) = 0. \tag{6.34}$$

(6.34) expresses the relationship between parameters $\mathbf{p}$ and the circuit performance of interest $\mathbf{f_p}$ in a general, implicit form.

3. Find the intersection points of the boundary with each parameter axis.
   (a) Augment equation (6.34) with parametric equations (6.40) of each parameter axis.
   (b) Solve the system of (6.34) and (6.40), the solution is the intersection of the parameter axis and the boundary manifold.
4. Use an analytical volume calculation formula to calculate the initial approximation to the yield – the hypervolume of the simplex defined by those intersections.
5. Volume refinement using additional boundary samples.
   (a) Choosing and computing additional points on the boundary manifold.
      i. Construct the manifold by connecting the points already found in previous refinements, and choose the centroid of this manifold to be the initial guess P_{guess} (see Fig. 6.16 for an example) for the point to be found on the boundary.
      ii. Either use the MPNR method to solve for the intersection point;
      iii. Or, use the line search method to solve for the intersection point.
         A. Construct the parametric equations for the line that goes through point P_{guess} and is perpendicular to the manifold constructed in step 5(a)i.
         B. Apply line search to find the intersection of the boundary and this line, using either the Newton–Raphson method or the secant/bisection method [43].
         C. If the Newton–Raphson method is used, the derivatives of function $h(\cdot)$ with respect to parameters $\mathbf{p}$ need to be evaluated by performing a sensitivity analysis of the circuit.
         D. If calculating the derivatives is computationally expensive or practically difficult, bisection, the secant method, and other line search techniques [47] can be applied to find the intersection points.
   (b) Update the yield estimate using the newly found boundary points.
      i. The yield volume increment is evaluated by calculating the hypervolume of the hyper-triangle defined by the newly found point and previous boundary points, using an analytical formula.
      ii. Based on the volume increment, an error metric is estimated for the current yield approximation.
      iii. If the error estimate is small enough, finish the algorithm;
      iv. Else, repeat the volume refinement procedure step 5 until sufficiently small error is obtained.

6.3.3   Implicit Formulation for the Boundary

We start by expressing the relationship between the performance metric and the varying parameters as an implicit *scalar equation* (6.34) which encapsulates all the complex nonlinear dynamics in the circuit. Suppose $\mathbf{p} \in \mathbb{R}^{N_p}$, then this scalar equation defines a $(N_p - 1)$-dimensional manifold/hypersurface in the parameter space. Our goal is to find points on the surface, i.e., to solve this nonlinear equation, efficiently and accurately.

Normally, there is only one acceptable region that is defined by:

$$h(\mathbf{p};\mathbf{f_p}) \equiv \mathbf{f_p}(\mathbf{p}) - \mathbf{f}_{\text{worst}} = 0. \tag{6.35}$$

In situations where there are several acceptable regions, YENSS can be applied on each acceptable region separately.

Depending on the problem or the circuit being considered, (6.34) can encapsulate different kinds of analyses. We next provide two examples to illustrate how (6.34) can be formulated: an SRAM read access time example which is based on SPICE-level transient simulation, and an SRAM static noise margin (SNM) example computed by a series of DC analyses of the circuit.

6.3.3.1   SRAM Read Access Time

Consider an SRAM cell with a read access time constraint. Recall from Sect. 6.1.2 that the read access time of a SRAM cell is the time between the word line going high and ΔBL reaching $\Delta BL_{\min}$.

Suppose the maximum acceptable read access time is t_f, then a direct application of (6.35) gives the boundary equation

$$h(\mathbf{p};\mathbf{f_p}) \equiv h(\mathbf{p};t_f) \equiv t(\mathbf{p})_{\Delta BL=\Delta BL_{\min}} - t_f = 0, \tag{6.36}$$

where $t(\mathbf{p})_{\Delta BL=\Delta BL_{\min}}$ denotes the time at which ΔBL reaches $\Delta BL_{\min}$.

Given a maximum read access time t_f, the voltage difference ΔBL at time $t = t_f$ can be calculated through performing a transient analysis on the SRAM circuit. If this voltage difference is smaller than the minimum voltage difference $\Delta BL_{\min}$ which can be detected by the sense amplifier, the read failure will occur. Therefore, we obtain another equation defining the boundary manifold

$$h(\mathbf{p};\mathbf{f_p}) \equiv h(\mathbf{p};\Delta BL) \equiv \Delta BL(\mathbf{p})_{t=t_f} - \Delta BL_{\min} = 0. \tag{6.37}$$

Note that (6.36) and (6.37) define the same boundary manifold, and that both of them are computed by performing a transient analysis. However, the formulation using (6.36) requires a transient simulation until the point when $\Delta BL = \Delta BL_{\min}$, while the formulation using (6.37) requires a transient simulation from $t = 0$ to $t = t_f$.

Besides, ΔBL can be directly obtained from the simulation results, while $t_{\Delta BL=\Delta BL_{min}}$ must be approximated in some way. Therefore, computing (6.37) is easier than (6.36), and we prefer the (6.37) formulation for this problem.

6.3.3.2   SRAM Static Noise Margin while Holding Data

The static noise margin of an SRAM cell (SNM) quantifies the amount of the voltage noise required at the internal nodes of the SRAM cell to flip the data stored. It is found graphically as the length of the side of the largest square which can fit between the voltage transfer curves (VTCs) of the two inverters. For example, as shown in Fig. 6.15, curves VTC1 and VTC2 represent the voltage transfer curves for the two inverters, respectively; and the square is the largest square that can be fitted between the two VTCs – the length of its side is the SNM.

Therefore, SNM is computed by two steps: (1) Perform a series of DC analysis on the two inverters in the SRAM cell, obtain the VTCs; (2) Draw the two VTCs together, with one of them inverted, and fit the largest square between the VTCs. The SNM is then computed to be the length of the side of this square.

If we denote SNM to be S and the minimum acceptable SNM to be S_{min}, then the boundary equation is

$$h(\mathbf{p}; \mathbf{f_p}) \equiv h(\mathbf{p}; S) \equiv S(\mathbf{p})_{t=t_f} - S_{min} = 0. \tag{6.38}$$

Again, we have a scalar equation to describe the boundary, and the computation of this scalar equation is via DC analysis and identification of the largest square that fits between the VTCs.

6.3.4   Solving for the Boundary Using Line Search

The main computational effort in YENSS is spent on solving for the boundary manifold. This includes step 3 (to find the intersection of the boundary and each parameter axis), and step 5 (to adaptively identify more points on the boundary). In

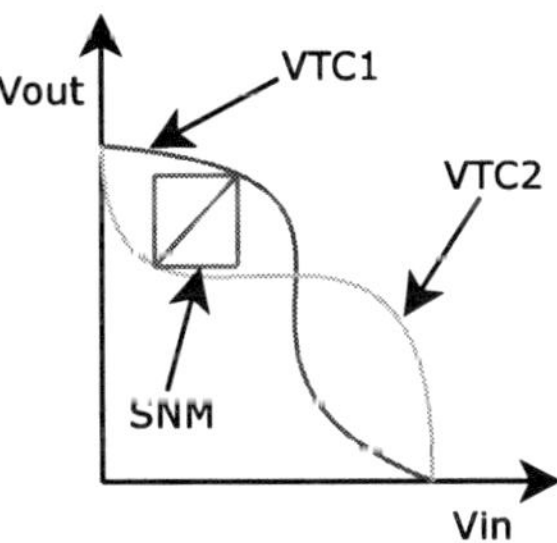

Fig. 6.15 Computing the static noise margin of an SRAM cell

this section, we discuss how a line search method is employed to solve for the boundary.

To start the YENSS algorithm, we need to first solve for the intersections of the boundary with each parameter axis. Note that it is possible that the boundary defined by the nonlinear scalar equation does not have any intersection with a parameter axis. However, in practice, parameter values are always bounded, hence the minimum/maximum points can be used for the intersection of the boundary with the parameter axis.

Solving for intersections of the boundary with each parameter axis is equivalent to solving the nonlinear equation (6.34) by fixing all the parameters but one (say p_i) at their nominal parameter values.

$$h(\mathbf{p}, \mathbf{f_p}) = 0$$
$$p_j = p_{j\text{nom}}, \quad \forall j \neq i$$

$$(6.39)$$

So the only unknown variable in the equation (6.34) is now a scalar p_i, rather than a vector $\mathbf{p}$ – we obtain a scalar equation with a scalar unknown variable. Therefore, we can apply any nonlinear solver, such as the secant method or the Newton–Raphson method, to find the intersections.

To use the Newton–Raphson method, however, we also need to calculate the Jacobian $\mathrm{d}h/\mathrm{d}\mathbf{p}$. Depending on the analysis used to compute $h(\mathbf{p}, \mathbf{f_p})$, we employ a corresponding sensitivity analysis to compute this quantity.

It is interesting to note that up to this stage, the computation is linear in the number of parameters. This implies that if the performance function is a linear or nearly linear function of the parameters then the computational expense of this method is linear in the number of parameters. Many practical problems have an approximately linear boundary surface, and therefore YENSS can be significantly faster than Monte-Carlo for these problems [9].

More generally, the earlier procedure to find the intersection points can be formulated as a line search method. Indeed, it is a special case of finding the intersections of the boundary manifold with any straight line – the line to be searched is the parameter axis in this case.

To force the solution of $h(\mathbf{p}, \mathbf{f_p}) = 0$ to lie on a line, we augment this equation by the equations defining the line. In a N_p-dimensional space, a straight line can be characterized by $N_p - 1$ linear equations. Equivalently, we can also use the N_p parametric equations

$$\mathbf{p} = \mathbf{p}_0 + s\mathbf{u} \tag{6.40}$$

as the line constraint. In (6.40), $\mathbf{p}_0$ is the coordinate of a point that is on the line, $\mathbf{u}$ the unit vector denoting the direction of the line, and s is a new scalar unknown variable introduced to parameterize the line.

For example, the parametric equation of p_1 axis can be written as:

$$p_1 = s,$$
$$p_i = 0, \quad 2 \le i \le N_p,$$
$$(6.41)$$

where $\mathbf{p}_0 = [0, \ldots, 0]^T$, and $\mathbf{u} = [1, 0, \ldots, 0]^T$.

Before augmenting with line constraint equations, we have a single scalar equation $h(\mathbf{p}, \mathbf{f_p}) = 0$, but the number of parameters $\mathbf{p}$ is N_p. Therefore, the number of equations is less than the number of unknowns, and the equation system is under-determined.

After the performance constraint equation $h(\mathbf{p}, \mathbf{f_p}) = 0$ is augmented by the line constraint equations, we have one performance constraint equation and N_p parametric equations for the line, and we have N_p unknown variables $\mathbf{p}$ and one variable s introduced to parameterize the line. Therefore, the number of unknown variables and the number of equations are both $N_p + 1$, and the augmented equations are no longer under-determined.

Therefore, the Newton–Raphson method can be applied to this square system to find the intersection points. Or equivalently, we can formulate the scalar equation for the scalar unknown variable s as

$$h(\mathbf{p}, \mathbf{f_p}) = h(\mathbf{p}_0 + s\mathbf{u}, \mathbf{f_p}) = 0,$$
$$(6.42)$$

so that simpler and more robust one-dimensional nonlinear solvers, such as the bisection or secant methods, can be applied to search the line.

After the intersections with the parameter axes are found, the line search procedure is performed iteratively to find additional points on the boundary. After the points have been found, we can construct locally linear $(N_p - 1)$-dimensional hyperplanes from sets of N_p points. Stitching together all those hyperplanes gives a piecewise linear approximation of the boundary.

The key problem here is to ensure that the points are a good representation of the boundary manifold, at least from the standpoint of hypervolume computation. To achieve this, one desired property is that the points on the boundary evenly spread out – this gives a better global approximation of the boundary than the case where many points are clustered in a small region.

To try to align all the points evenly on the boundary surface, we construct a specific line to be searched with – it has two properties: (1) it goes through the centroid P_{guess} of the N_p points that constitute a locally linear hyperplane; (2) it is perpendicular to this locally linear hyperplane. Constructing this line and using P_{guess} as an initial guess or initial search point, we apply the line search method on the augmented system to find the intersection of this line with the real boundary. If the real boundary is almost linear, this line constraint will lead to a new point that is almost at the centroid of the boundary. In this case, the initial guess P_{guess} is also very close to the real solution, which implies that the nonlinear solver will converge in very few iterations.

For illustration, the results of two boundary refinement iterations in a 2-dimensional parameter space are shown in Fig. 6.16a, b, where the bold curve represents the boundary curve defined by (6.35). In Fig. 6.16a, P_{01}, P_{02} are two intersections with parameter axes, calculated in the first step. Then $P_{1\text{guess}}$ is chosen to be the

Fig. 6.16 Finding points on the boundary curve using line search (two iterations)

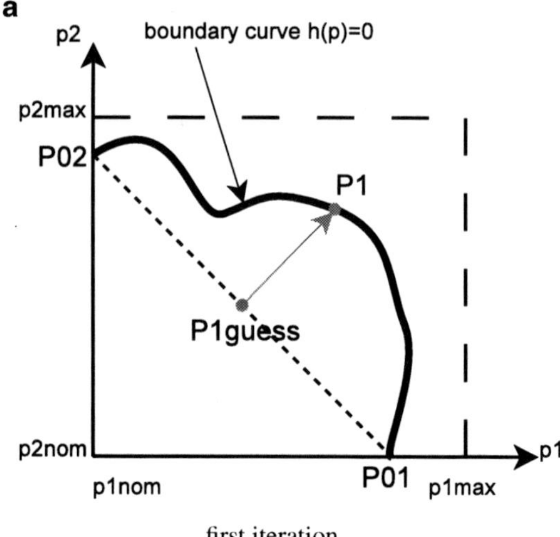

first iteration

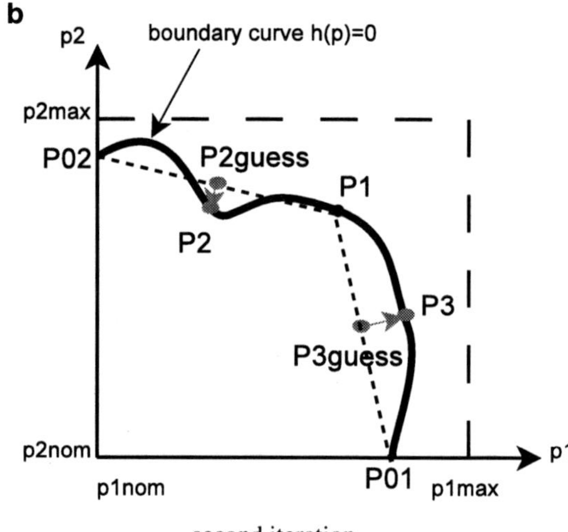

second iteration

centroid of $P_{01}P_{02}$, and the line $P_{1guess} P_1$ which is perpendicular to $P_{01}P_{02}$ is constructed. The performance constraint equation $h(\mathbf{p}, \mathbf{f}_p) = 0$ is then augmented by the parametric equations for line $P_{1guess} P_1$. Using P_{1guess} to be the initial guess for Newton–Raphson solver, it finally converges to P_1, and Newton iterations effectively search along the line $P_{1guess} P_1$.

In the second boundary refinement iteration (shown in Fig. 6.16b), P_{2guess} and P_{3guess} are selected to be the centroid of $P_{02}P_1$ and $P_{01}P_1$, respectively, and Newton–Raphson will follow the lines $P_{2guess} P_2$ and $P_{3guess} P_3$, converging to P_2 and P_3, respectively. Also notice that the search distance for P_2 and P_3 is much less than that for P_1. This is another good feature of this method – as the boundary becomes more linear, the search distance decreases and therefore the computational cost is lessened.

Repeating this procedure, we will finally obtain enough points on the boundary curve needed for yield estimation. Importantly, it is straightforward to apply this procedure to high-dimensional problems.

6.3.4.1 Dealing with Multiple Performance Constraints

The line search method discussed earlier is directly applicable to the problem with one performance constraint (i.e., there is only one performance constraint equation $h(\mathbf{p}, \mathbf{f_p}) = 0$). When there are multiple performance constraints, however, we have to find the common interior of boundaries corresponding to N_p performance constraint equations $\mathbf{h}(\mathbf{p}, \mathbf{f_p}) = [h_1(\mathbf{p}, \mathbf{f_p}), \ldots, h_{N_p}(\mathbf{p}, \mathbf{f_p})] = 0$.

Therefore, to find the boundary of the intersection of acceptable regions, we must find the "first" point among the points the line intersects, with boundaries corresponding to different constraints. For example, in Fig. 6.17, suppose the dotted line is the line to be searched. On this line, there are three intersection points, with the three curves which represent three different constraints on the circuit perfor mance. Among the three intersection points, the "first" intersection point is A.

To do this, the previous line search method needs to be adjusted. Suppose there are N_p performance constraints, defined by $h_i(\mathbf{p}) = 0$, $(i = 1, \ldots, N_p)$. The algorithm to find the samples on the interior of all the constraint boundaries is:

1. Mark all the constraints to be unsatisfied;
2. Loop until all the constraints are satisfied;
    (a) Pick up one unsatisfied constraint, calculate the intersection point using the line search method;
    (b) Test all the unsatisfied constraints;
    (c) If all the constraints are met, terminate;
    (d) Else, go to step 2a.

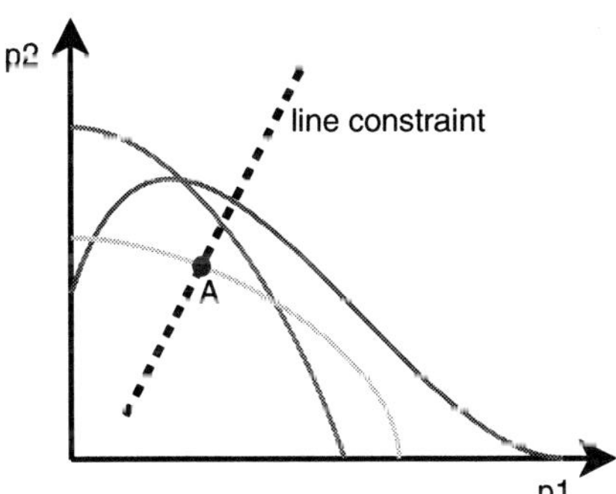

Fig. 6.17 Illustration of line search when there are multiple constraints

6.3.5 Solving for the Boundary Using Moore–Penrose Pseudo-Inverse Newton–Raphson

Another option is to use the Moore–Penrose pseudo-inverse Newton Raphson (MPNR) method [46] to solve the scalar equation (6.35) directly, without augmenting with additional linear equations. While the classical Newton–Raphson method can only be applied to square systems, MPNR can be used to solve underdetermined equations.

The algorithm is similar to the classical Newton–Raphson method: to solve $f(x) = 0$, it starts with an initial guess x_0, and then updates x_i iteratively until the accuracy is within tolerance.

Its difference from classical Newton–Raphson is in the update step. In classical Newton–Raphson, the update step is

$$x_{i+1} = x_i - J(x_i)^{-1} f(x_i), \tag{6.43}$$

where $J(x_i)$ is the Jacobian matrix at $x = x_i$. However, in our problem, we have 1 scalar equation and N_p unknown variables; the size of the Jacobian matrix $J(x_i)$ is $1 \times N_p$ (not square), and hence its inverse does not exist.

Instead of using the inverse of $J(x_i)$, MPNR uses its Moore–Penrose pseudo-inverse

$$J(x_i)^+ = J(x_i)^T [J(x_i) J(x_i)^T]^{-1} \tag{6.44}$$

in the update step. Therefore, the update step formula in MPNR is

$$x_{i+1} = x_i - J(x_i)^T [J(x_i) J(x_i)^T]^{-1} f(x_i). \tag{6.45}$$

Unlike line search, which has a unique solution, MPNR may end up with any solution to the nonlinear equation. However, it can be proved [46] that, under the right circumstances, MPNR will converge to a point on the solution curve that is closest to the initial guess. Therefore, if we choose the initial guess in the same way as in the line search method, MPNR will also generate samples that are well spaced on the boundary.

6.3.6 Calculating the Jacobian Matrix Using Sensitivity Analysis

In order to use the Newton–Raphson method in line search, or the MPNR method, the Jacobian matrix of the system must be calculated by performing a sensitivity analysis of the circuit. Given (6.35), we differentiate $h(\,\cdot\,)$ with respect to $\mathbf{p}$:

$$\frac{dh}{d\mathbf{p}} = \frac{\partial h}{\partial \mathbf{p}} + \frac{\partial h}{\partial \mathbf{f_p}} \frac{\partial \mathbf{f_p}}{\partial \mathbf{p}}. \tag{6.46}$$

In (6.46), $\partial h/\partial \mathbf{p}$ and $\partial h/\partial \mathbf{f_p}$ are available after the function $h(\mathbf{p}, \mathbf{f_p})$ is defined. However, we usually do not have an analytical form for the performance metrics $\mathbf{f_p}(\mathbf{p})$, the quantity $\partial \mathbf{f_p}/\partial \mathbf{p}$ remains unknown, and must be evaluated by performing a simulation.

Depending on the type of analysis used, the algorithm to compute the sensitivity $\partial \mathbf{f_p}/\partial \mathbf{p}$ varies. We now present derivations of sensitivity analysis based on transient simulation, which is extensively used in many applications.

6.3.6.1   Transient Simulation Based Sensitivity Evaluation

In many problems such as the SRAM read access failure problem, evaluation of the boundary equation is performed through a transient analysis of the circuit, i.e., by doing a numerical integration on the differential algebraic equations of the circuit

$$\frac{d}{dt}\mathbf{q}(\mathbf{x}) + \mathbf{f}(\mathbf{x}) + \mathbf{b}(t) = 0, \tag{6.47}$$

where $\mathbf{x}$ are the unknowns (node voltages and branch currents), and $\mathbf{b}(t)$ are the inputs to the circuit.

Again taking the SRAM read access failure problem as an example, we have the boundary equation:

$$h(\mathbf{p}) \equiv \Delta BL - \Delta BL_{min} = \mathbf{c}^T \mathbf{x}(t_f; \mathbf{p}) - \Delta BL_{min} = 0, \tag{6.48}$$

where $\mathbf{c}^T$ is a vector which extracts the voltage difference $\mathbf{f_p} = \Delta BL$ from the nodal voltages $\mathbf{x}$.

Therefore, the derivatives of $\mathbf{f_p}(\mathbf{p})$ with respect to $\mathbf{p}$ is

$$\frac{\partial \mathbf{f_p}(\mathbf{p})}{\partial \mathbf{p}} = \mathbf{c}^T \frac{\partial \mathbf{x}}{\partial \mathbf{p}}. \tag{6.49}$$

In order to calculate $\partial \mathbf{x}/\partial \mathbf{p}$, we differentiate the circuit equations (6.47) with respect to the parameters $\mathbf{p}$:

$$\frac{d}{d\mathbf{p}}\left[\frac{d}{dt}\mathbf{q}(\mathbf{x}) + \mathbf{f}(\mathbf{x}) + \mathbf{b}(t)\right] = 0. \tag{6.50}$$

Explicitly, we obtain

$$\frac{d}{dt}\left[\frac{\partial \mathbf{q}}{\partial \mathbf{x}}\frac{\partial \mathbf{x}}{\partial \mathbf{p}} + \frac{\partial \mathbf{q}}{\partial \mathbf{p}}\right] + \frac{\partial \mathbf{f}}{\partial \mathbf{x}}\frac{\partial \mathbf{x}}{\partial \mathbf{p}} + \frac{\partial \mathbf{f}}{\partial \mathbf{p}} + \frac{d\mathbf{b}}{d\mathbf{p}} = 0, \tag{6.51}$$

where $\partial \mathbf{q}/\partial \mathbf{x}$ and $\partial \mathbf{f}/\partial \mathbf{x}$ are already calculated during the transient simulation, and $\partial \mathbf{q}/\partial \mathbf{p}$, $\partial \mathbf{f}/\partial \mathbf{p}$, $d\mathbf{b}/d\mathbf{p}$ can be computed.

Therefore, (6.51) is a set of differential equations with unknowns $\partial \mathbf{x}/\partial \mathbf{p}$, and hence can be solved using any numerical integration method, such as the Backward-Euler method or the trapezoidal method [52, 53].

Thus, the sensitivities of the nodal voltages and branch currents $\mathbf{x}$ with respect to varying parameters $\mathbf{p}$ can be calculated, following which $\partial \mathbf{h}/\partial \mathbf{p}$ can be calculated.

6.3.7 Adaptive Hypervolume Refinement and Error Estimation

6.3.7.1 Analytical Formula to Compute the Hypervolume

Given non-collinear $N + 1$ points in a N-dimensional space, we can construct a simplex. For example, three points in a 2-dimensional plane constitute a triangle. The area/volume/hypervolume of this simplex can be simply calculated by:

$$
C_{\text{simplex}} = \frac{1}{N!} \begin{vmatrix} x_{1,1} & x_{1,2} & \cdots & x_{1,N} & 1 \\ x_{2,1} & x_{2,2} & \cdots & x_{2,N} & 1 \\ \vdots & \vdots & \vdots & \vdots & \vdots \\ x_{N+1,1} & x_{N+1,2} & \cdots & x_{N+1,N} & 1 \end{vmatrix},
\tag{6.52}
$$

where $(x_{i,1}, \ldots, x_{i,N})$ are the coordinates of the ith point. Notice that this formula actually might lead to a "negative hypervolume" – it can be shown that the sign represents the "direction" of the simplex, and can be utilized in the volume refinement step, as discussed later.

6.3.7.2 Yield Hypervolume Calculation and Error Control

As the new points on the boundary manifold are found, we update the hypervolume estimate by adding/subtracting the hypervolume of the new simplexes we find. For example, two iterations of the refinement and hypervolume update procedure are shown in Fig. 6.18. S_0 is the original area estimate after the axis intersections have been found. After one iteration, P_1 is found, and S_1 (area of triangle $P_{01}P_{02}P_1$) is added to S_0 to refine the area estimate. After two iterations, P_2 and P_3 are found, and S_2 (area of triangle $P_{02}P_1P_2$) is subtracted and S_3 (area of triangle $P_{01}P_1P_3$) is added. So, after two level of refinements, the yield volume is $S_0 + S_1 - S_2 + S_3$. It is clear that after several iterations, we can get a fairly good estimate of the total area under the boundary curve.

Also note that the computational cost for calculating the determinant (6.52) is expensive ($O(N^3)$). However, since the points that constitute the new simplex share all but one points with the vertices of the existing simplex, we can reduce the computational cost to $O(N)$. For example, in Fig. 6.18 triangle $P_{01}P_{02}P_1$ and triangle $P_{01}P_1P_3$ share the edge $P_{01}P_1$. Therefore, the ratio of S_1 to S_3 is just the ratio of the distance from P_{02} to $P_{01}P_1$ and the distance from P_3 to $P_{01}P_1$, which can

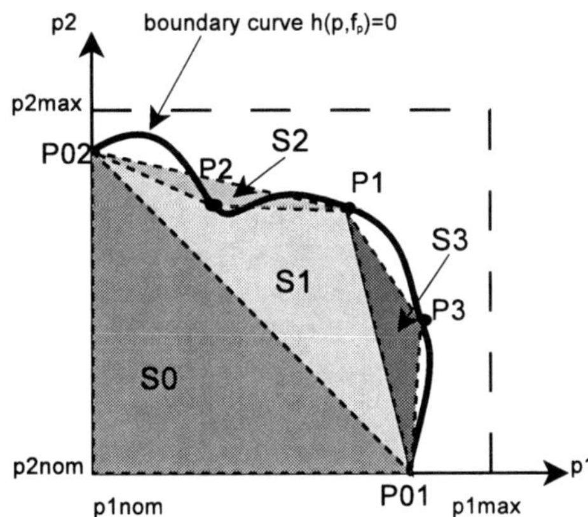

Fig. 6.18 Illustration of yield volume update scheme

be computed in $O(N)$ time since the normal vector is already known. Similarly, this applies to the computation of higher-dimensional hypervolumes.

The relative incremental hypervolume computed also provides an estimate of the error from the true yield. Based on the volume increment, YENSS adaptively decides whether to continue further iterations, depending on the accuracy desired. For example, in Fig. 6.18, if $S2$ is small enough (smaller than some predefined threshold), there is no need to find more points on curve $P_{02} - P_1$; but if $S3$ is still large, we can continue the curve-finding scheme to find more points on curve $P_1 - P_{01}$, until the designated accuracy is reached.

This automatic error control scheme helps to avoid redundant simulations that increase computational cost unnecessarily. If the boundary is a straight line, the error calculated by this scheme after one refinement is 0, and the algorithm terminates with perfect accuracy. This explains how this error estimation works and why YENSS is extremely efficient if the performance function varies linearly with parameter variations.

6.4   Examples and Comparisons

To illustrate how YENSS works, we first apply it on an illustrative multiple-constraint problem. Then we apply both YENSS and the Monte-Carlo method to the SRAM read access failure problem mentioned previously.

6.4.1   An Illustrative Example of YENSS

This example shows how YENSS works on a multiple-constraint problem. In this problem, we have two independent parameters p_1 and p_2, which are uniformly distributed on the interval [0, 0. 5]. There are three constraints on the parameters:

$$2p_1 + p_2 - 1 \le 0,$$
$$p_1 + 2p_2 - 1 \le 0,$$
$$1.05^2 - (p_1 - 1)^2 - (p_2 - 1)^2 \le 0. \tag{6.53}$$

We would like to compute the parametric yield.

Figure 6.19 shows the boundaries corresponding to the three constraints. From the previous sections, we know that the yield is the ratio of the area inside all the boundaries to the area of the box $[0, 0.5] \times [0, 0.5]$.

Applying YENSS, we first identify points on the interior of the three boundaries. As shown in Fig. 6.19, the dots are the points found by applying the line search method. After 6 refinements, 31 points are found that approximate the boundary. As can be seen, they approximate the boundary very well, and are almost evenly spaced on the boundary. The area inside boundaries is computed to be 0.1388, and therefore, we have the yield estimate to be $\frac{0.1388}{0.25} = 55.52\%$.

6.4.2 Application to SRAM Read Access Failure

We now apply YENSS and the Monte-Carlo method on the SRAM example mentioned previously. Considering the threshold voltages of $M1$ and $M2$ to be the varying parameters of interest, we obtain the scalar equation (6.34) for the boundary to be

$$h(V_{\text{th1}}, V_{\text{th2}}) \equiv (\mathbf{c}_{BL}^T - \mathbf{c}_{BL\_B}^T)\Phi(T_{\max}; V_{\text{th1}}, V_{\text{th2}}) - \Delta BL_{\min} = 0, \tag{6.54}$$

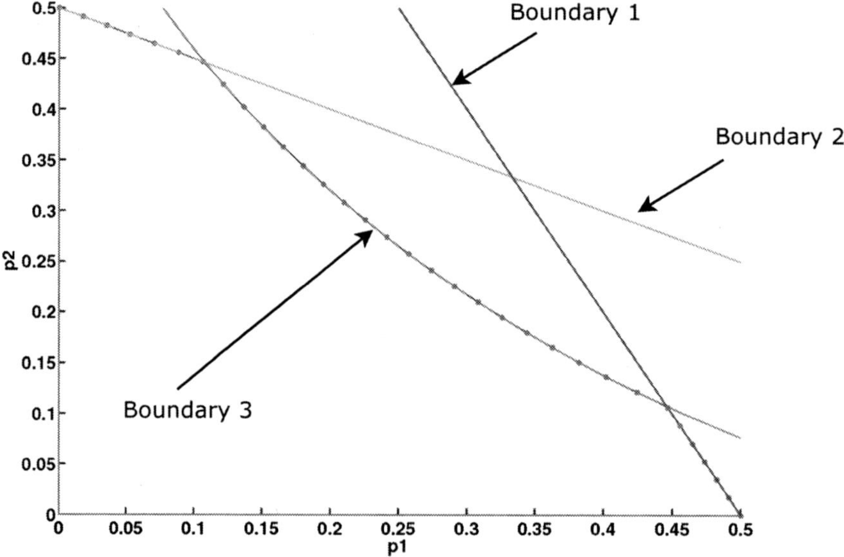

Fig. 6.19 Illustration of YENSS on a multiple-constraint problem

where: $\Phi(\cdot)$ is the state-transition function defining the node voltages and branch currents in the circuit, which is calculated by performing a transient simulation; $(\mathbf{c}_{BL}^T - \mathbf{c}_{BL\_B}^T)$ is the vector which selects the voltage difference between node BL and BL_B; and $T_{\max}$ is the specified maximum read access time, defined as the time required to produce the minimum voltage difference $\Delta BL_{\min}$ between BL and BL_B that can be detected by the sense amplifier (normally defined by $\Delta BL_{\min} \simeq 0.1Vdd$).

Assume that the threshold voltages are independent and are uniformly distributed on $[0.1V, 0.7V]$, and that $\Delta BL_{\min} = 168$mV at $T_{\max} = 3.508$ ns. We first vary the threshold voltage of M1 (V_{th1}) and M2 (V_{th2}), and run simulations with different (V_{th1}, V_{th2}) pairs. Figure 6.20a depicts $h(V_{th1}, V_{th2})$ (i.e., $\Delta BL - \Delta BL_{\min}$) surface over ($V_{th1}$, V_{th2}) parameter space. From this surface generated by brute-force simulations, the boundary curve can be identified, visually as shown in Fig. 6.20b. There are two problems with the brute-force method: (1) the accuracy is limited by the step size on each parameter axis – each point generated in this method does not lie exactly on the boundary curve, and this is obviously observed in Fig. 6.20b; (2) the number of simulations can be too large to be affordable. Although some heuristics can be applied, (for example, one may safely skip simulating the area where V_{th1} and V_{th2} are both small), the computational cost is still expensive.

We then apply YENSS to find the boundary curve directly. Figure 6.21 shows the results of YENSS using different levels of refinements, as well as results from Monte-Carlo simulation. In Fig. 6.21a, only three levels of refinements are performed – seven points (marked on the curve) on the boundary curve are obtained. The boundary curve obtained is still coarse and jagged of this level of refinement. Figure 6.21b, and c shows the results of 4 and 5 levels of refinements, by which point a smooth boundary curve is obtained. The boundary found by YENSS is validated by Monte-Carlo simulation, as shown in Fig. 6.21d.

Detailed comparisons regarding accuracy, simulation time and speed-ups between YENSS and Monte-Carlo is shown in Table 6.1. To achieve an accuracy of 1% with 95% confidence, Monte-Carlo needs over 6,000 simulations. However, only three levels of refinement using YENSS, even though the curve is still jagged (shown in Fig. 6.21a), gives an accuracy of 1%, with 255 $\times$ speedup over the Monte-Carlo method.

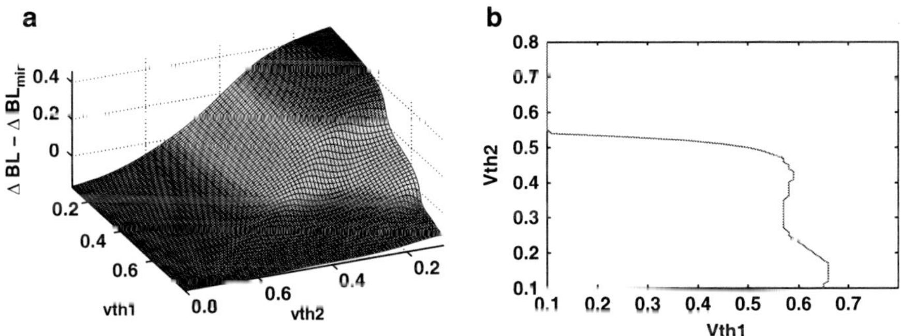

Fig. 6.20 Brute force simulation to get the boundary curve

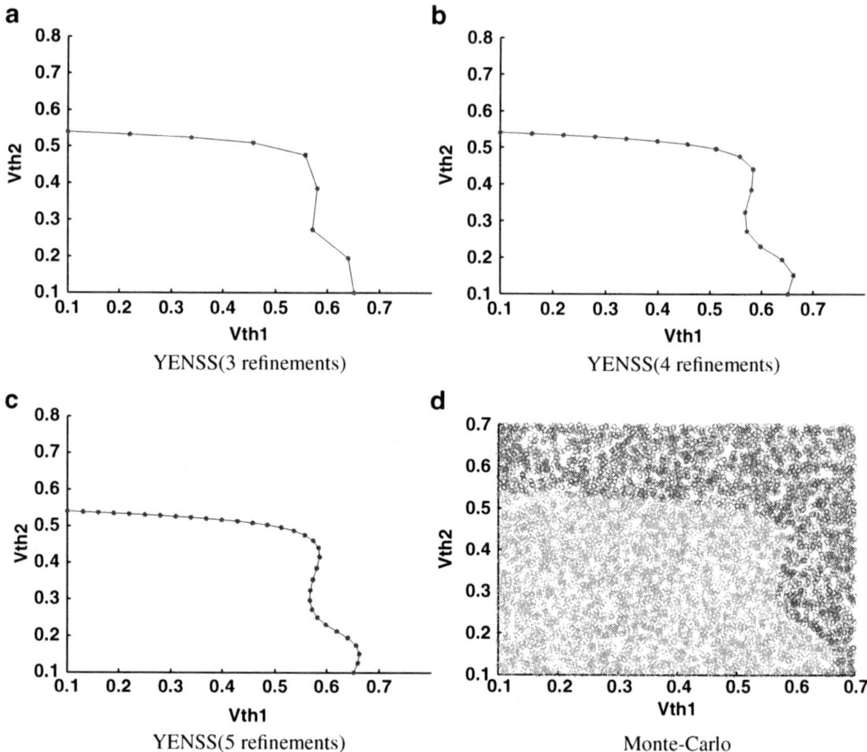

Fig. 6.21 Simulation results of YENSS and Monte-Carlo on SRAM read access time constraint

Table 6.1 Comparison of YENSS vs. Monte-Carlo for SRAM yield

Method	Accuracy	Yield	Number of transient simulations	Speedup
Monte-Carlo	10%	0.6173	80	
Monte-Carlo	1%	0.5730	6640	1 ×
YENSS				
3 levels	1%	0.5756	26	255 ×
4 levels	0.1%	0.5789	41	162 ×
5 levels	< 0.1%	0.5797	66	101 ×

6.5 Summary

In this chapter, we have formulated the yield estimation problem as a probabilistic hypervolume computation problem. We classified existing methods into two classes: statistical and deterministic/mixed methods, and have reviewed available algorithms. We have discussed the deterministic YENSS algorithm in detail, illustrated its application to an SRAM circuit, and provided comparisons against the classical Monte-Carlo method.

References

1. Nassif SR (2001) Modeling and analysis of manufacturing variations. In Proceedings of the IEEE conference on custom integrated circuits, 6–9 May 2001, pp 223–228
2. Papoulis A (1984) Probability, random variables, and stochastic processes. Mc-Graw Hill, New York
3. The International Technology Roadmap for Semiconductors (2008) http://www.itrs.net/
4. Maly W, Heineken H, Khare J, Nag PK (1996) Design for manufacturability in submicron domain. In Proceedings of the IEEE/ACM international conference on computer-aided design ICCAD-96. Digest of Technical Papers, 10–14 November 1996, pp 690–697
5. Gupta P, Kahng AB (2003) Manufacturing-aware physical design. In Proceedings of the ICCAD-2003 computer aided design international conference, 9–13 November 2003, pp 681–687
6. Heald R, Wang P (2004) Variability in sub-100nm sram designs. In Proceedings of the ICCAD-2004 computer aided design IEEE/ACM international conference, 7–11 November 2004, pp 347–352
7. Rabaey JM, Chandrakasan A, Nikolic B (2003) Digital integrated circuits, 2nd edn. (Printice Hall Electronics and Vlsi Series). Prentice Hall, Englewood Cliffs, NJ, USA
8. Agarwal K, Nassif S (2007) Characterizing process variation in nanometer cmos. In Proceedings of the 44th ACM/IEEE design automation conference DAC '07, 4–8 June 2007, pp 396–399
9. Chenjie Gu, Roychowdhury J (2008) An efficient, fully nonlinear, variability-aware non-monte-carlo yield estimation procedure with applications to sram cells and ring oscillators. In Proceedings of the Asia and South Pacific design automation conference ASPDAC 2008, 21–24 March 2008, pp 754–761
10. Singhal K, Pinel J (1981) Statistical design centering and tolerancing using parametric sampling. Circuits Syst, IEEE Trans on 28(7):692–702
11. Wojciechowski JM, Vlach J (1993) Ellipsoidal method for design centering and yield estimation. Comput Aided Des Integr Circuits Syst, IEEE Trans on 12(10):1570–1579
12. Director S, Hachtel G (1977) The simplicial approximation approach to design centering. Circuits Syst, IEEE Trans on 24(7):363–372
13. Maly W, Director SW (1980) Dimension reduction procedure for the simplicial approximation approach to design centering. IEE Proc G Electronic Circuits Syst 127(6):255–259
14. Antreich K, Koblitz R (1982) Design centering by yield prediction. IEEE Trans Circuits and Syst 29(2):88–96
15. Seifi A, Ponnambalam K, Vlach J (1999) A unified approach to statistical design centering of integrated circuits with correlated parameters. IEEE Trans Circuits Syst I, Fundam Theory Appl 46(1):190–196
16. Low KK, Director SW (1989) A new methodology for the design centering of ic fabrication processes. In Proceedings of the IEEE International Conference on Computer-Aided Design ICCAD-89. Digest of Technical Papers, 5–9 November 1989, pp 194–197
17. Polak E, Sangiovanni-Vincentelli A (1979) Theoretical and computational aspects of the optimal design centering, tolerancing, and tuning problem. IEEE Trans Circuits Syst 26(9):795–813
18. Brayton R, Director S, Hachtel G (1980) Yield maximization and worst-case design with arbitrary statistical distributions. Circuits Syst, IEEE Trans, 27(9):756–764
19. Chopra K, Shah S, Srivastava A, Blaauw D, Sylvester D (2005) Parametric yield maximization using gate sizing based on efficient statistical power and delay gradient computation. In Proceedings of the ICCAD-2005 computer-aided design IEEE/ACM international conference, 6–10 November 2005, pp 1023–1028
20. Director SW, Feldmann P, Krishna K (1992) Optimization of parametric yield: a tutorial. In Proceedings of the custom integrated circuits conference the IEEE 1992, May 3–6, 1992, pp 3.1.1–3.1.8

21. Stehr G, Graeb H, Antreich K (2003) Performance trade-off analysis of analog circuits by normal-boundary intersection. In Proceedings of the Design Automation Conference, 2003, 2–6 June 2003, pp 958–963
22. Toumazou C, Moschytz GS, Gilbert B (eds) (2002) Trade-Offs in analog circuit design: the designer's companion. Kluwer Academic Publishers, Dordrecht (Hingham, MA)
23. Eeckelaert T, McConaghy T, Gielen G (2005) Efficient multiobjective synthesis of analog circuits using hierarchical pareto-optimal performance hypersurfaces. In Proceedings of the design, automation and test in Europe, 07–11 March 2005, pp 1070–1075
24. Mukhopadhyay S, Mahmoodi H, Roy K (2005) Modeling of failure probability and statistical design of SRAM array for yield enhancement in nanoscaled CMOS. Comput -Aided Des Integr Circuits Syst, IEEE Trans 24(12):1859–1880
25. Swidzinski JF, Chang K (2000) Nonlinear statistical modeling and yield estimation technique for use in Monte Carlo simulations [Microwave Devices and ICs]. Microwave Theory Tech, IEEE Trans 48(12):2316–2324
26. Liu JS (2002) Monte Carlo strategies in scientific computing. Springer, Berlin
27. Robert CP, Casella G (2005) Monte Carlo statistical methods (Springer texts in statistics). Springer-Verlag New York, Inc., Secaucus, NJ, USA
28. Kundert KS (1995) The designer's guide to spice and spectre. Kluwer Academic Publishers, Dordrecht (Hingham, MA)
29. Kundert K, White J, Sangiovanni-Vincentelli A (1990) Steady-state methods for simulating analog and microwave circuits. Kluwer Academic Publishers, Dordrecht (Hingham, MA)
30. Seevinck E, List FJ, Lohstroh J (1987) Static-noise margin analysis of MOS SRAM cells. IEEE J Solid-State Circuits 22(5):748–754
31. Devgan A, Kashyap C (2003) Block-based static timing analysis with uncertainty. In Proceedings of the ICCAD-2003 computer aided design international conference, 9–13 November 2003, pp 607–614
32. Chang H, Sapatnekar SS (2003) Statistical timing analysis considering spatial correlations using a single pert-like traversal. In ICCAD '03: Proceedings of the 2003 IEEE/ACM international conference on Computer-aided design, Washington, DC, USA, November 2003. IEEE Computer Society, p 621
33. Kanj R, Joshi R, Nassif S (2006) Mixture importance sampling and its application to the analysis of sram designs in the presence of rare failure events. In Proceedings of the 43rd ACM/IEEE design automation conference, 24–28 July 2006, pp 69–72
34. Hastings WK (1970) Monte carlo sampling methods using markov chains and their applications. Biometrika 57(1):97–109
35. Keramat M, Kielbasa R (1997) A study of stratified sampling in variance reduction techniques for parametric yield estimation. In Proceedings of the IEEE international symposium on circuits and systems ISCAS '97, vol3, 9–12 June 1997, pp 1652–1655
36. Gallaher LJ (1973) A multidimensional monte carlo quadrature with adaptive stratified sampling. Commun ACM 16(1):49–50
37. Stein M (1987) Large sample properties of simulations using latin hypercube sampling. Technometrics 29(2):143–151
38. Hammersley JM, Morton KW (1956) A new Monte Carlo technique: antithetic variates. In Proceedings of the Cambridge Philosophical Society, vol52 of Proceedings of the Cambridge Philosophical Society, July 1956, pp 449–475
39. Director S, Hachtel G (1977) The simplicial approximation approach to design centering. Circuits Syst, IEEE Trans 24(7):363–372
40. Director S, Hachtel G, Vidigal L (1978) Computationally efficient yield estimation procedures based on simplicial approximation. Circuits Syst, IEEE Trans 25(3):121–130
41. Biernacki RM, Bandler JW, Song J, Zhang QJ (1989) Efficient quadratic approximation for statistical design. IEEE Trans Circuit Syst 36(11):1449–1454
42. Antreich KJ, Graeb HE, Wieser CU (1994) Circuit analysis and optimization driven by worst-case distances. Comput Aided Des Integr Circuits Syst IEEE Trans 13(1):57–71, Jan. 1994.

43. Heath MT (1996) Scientific computing: an introductory survey. McGraw-Hill Higher Education, New York
44. Abdel-Malek HL, Hassan AKSO (1991) The ellipsoidal technique for design centering and region approximation. Comput Aided Des 10(8):1006–1014
45. Srivastava S, Roychowdhury J (2007) Rapid estimation of the probability of SRAM failure due to MOS threshold variations. In Custom integrated circuits conference, 2007., Proceedings of the IEEE 2007, September 2007
46. Allgower EL, Georg K (1990) Numerical continuation methods. Springer-Verlag, New York
47. Nocedal J, Wright SJ (1999) Numerical optimization. Springer, New York
48. Li P (2006) Statistical sampling-based parametric analysis of power grids. Comput -Aided Des Integr Circuits Syst, IEEE Trans 25(12):2852–2867
49. Chang H, Zolotov V, Narayan S, Visweswariah C (2005) Parameterized block-based statistical timing analysis with non-Gaussian parameters, nonlinear delay functions. In Proceedings of the 42nd Design Automation Conference, 2005, 13–17 June 2005, pp 71–76
50. Michael C, Ismail MI (1993) Statistical modeling for computer-aided design of MOS VLSI circuits. Springer, Berlin
51. Golub GH, VanLoan CF (1996) Matrix computations (Johns Hopkins Studies in Mathematical Sciences). The Johns Hopkins University Press, Baltimore, MD, USA
52. Chua LO, Lin P-M (1975) Computer-aided analysis of electronic circuits : algorithms and computational techniques. Prentice-Hall, Englewood Cliffs, NJ
53. Gear CW (1971) Numerical initial value problems in ordinary differential equations. Prentice-Hall series in automatic computation. Prentice-Hall, Englewood Cliffs, NJ

Chapter 7
Most Probable Point-Based Methods

Xiaoping Du, Wei Chen, and Yu Wang

7.1 Introduction

Most Probable Point (MPP) based methods are widely used for engineering reliability analysis and reliability-based design. Their major advantage is the good balance between accuracy and efficiency. The most commonly used MPP-based method is the First-Order Reliability Method (FORM) [14]. We will primarily discuss the FORM in this chapter.

Loosely speaking, reliability is the probability of success. In the MPP-based methods, the state of success is specified by computational models. The computational model is also called a limit-state function. A limit-state function is given by

$$Y = g(\mathbf{X}).\tag{7.1}$$

The model input contains random variables $\mathbf{X} = (X_1, \ldots, X_n)$, which are referred to as basic random variables because it is assumed that their joint probability density function (PDF) $f_X(\mathbf{x})$ is known. The model output Y is a performance characteristic. The state, either success (safety) or failure, can be determined by the value of Y. Without losing generality, we assume that when $Y = g(\mathbf{X}) > 0$, the state is success, and that when $Y = g(\mathbf{X}) \leq 0$, the state is failure. The reliability R, therefore, is the probability of $Y = g(\mathbf{X}) > 0$ and is given by

$$R = \Pr\{g(\mathbf{X}) > 0\},\tag{7.2}$$

where $\Pr\{\cdot\}$ denotes a probability.

The probability of failure p_f is defined by

$$p_f = \Pr\{g(\mathbf{X}) \leq 0\}.\tag{7.3}$$

X. Du (✉)
Department of Mechanical and Aerospace Engineering, Missouri University of Science and Technology, Rolla, MO, USA
e-mail: dux@mst.edu

A. Singhee and R.A. Rutenbar (eds.), *Extreme Statistics in Nanoscale Memory Design,*
Integrated Circuits and Systems, DOI 10.1007/978-1-4419-6606-3_7,
© Springer Science+Business Media, LLC 2010

R and p_f add up to 1.0; namely,

$$R + p_f = 1. \tag{7.4}$$

The limit-state function at the limit state $g(\mathbf{X}) = 0$ divides the input space, or the X-space, into two regions: safety region $\{\mathbf{X}|g(\mathbf{X}) > 0\}$ and failure region $\{\mathbf{X}|g(\mathbf{X}) \leq 0\}$, which are illustrated in Fig. 7.1.

Given the joint density of $\mathbf{X}, f_X(\mathbf{x})$, the probability of failure can be theoretically evaluated by

$$p_f = \Pr\{g(\mathbf{X}) \leq 0\} = \int_{g(\mathbf{x}) \leq 0} f_X(\mathbf{x}) \mathrm{d}\mathbf{x}. \tag{7.5}$$

For a two-dimensional limit-state function, the above equation becomes

$$p_f = \Pr\{g(X_1, X_2) \leq 0\} = \iint_{g(x_1, x_2) \leq 0} f_{X_1 X_2}(x_1, x_2) \mathrm{d}x_1 \mathrm{d}x_2. \tag{7.6}$$

For an n-dimensional limit-state function, the probability integral will be n-fold. In general, the multifold probability integral of $f_X(\mathbf{x})$ is over the failure region with a nonlinear boundary $g(\mathbf{X}) = 0$. As a result, it is difficult to compute the probability integral analytically. For most of engineering applications, only approximation methods are used to evaluate the probability integral. MPP-based methods are among those approximation methods.

The central idea of the MPP-based methods is to simplify the integrand $f_X(\mathbf{x})$ and to approximate the integration boundary $g(\mathbf{X}) = 0$. With the simplification and approximation, the probability of failure will be easily evaluated. When the integration boundary $g(\mathbf{X}) = 0$ is approximated by a linear function, the method becomes the FORM; likewise, if the boundary is approximated by a quadratic function, the method is called the Second-Order Reliability Method (SORM). In this chapter, we primarily discuss the FORM, starting from linear limit-state functions with normally distributed random variables and then extending the result to nonlinear limit-state functions. We then provide discussions on the SORM and some highlights of other MPP-based methods.

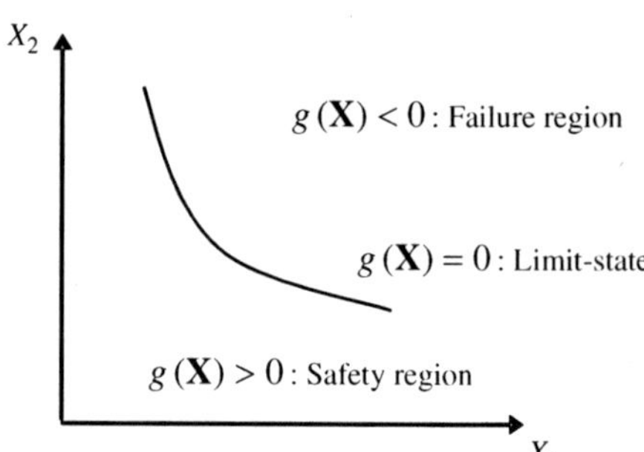

Fig. 7.1 Limit-state function

7.2 Linear Limit-State Functions with Normally Distributed Random Variables

Suppose $g(\mathbf{X})$ is a linear function and is given by

$$g(\mathbf{X}) = a_0 + a_1 X_1 + \cdots + a_n X_n = a_0 + \sum_{i=1}^{n} a_i X_i, \qquad (7.7)$$

where $X_i (i = 1, \ldots, n)$ are independently and normally distributed, and $a_i (i = 0, \ldots, n)$ are constants. If the mean and standard deviation of X_i are μ_i and σ_i, respectively, X_i can be transformed into a standard normal variable U_i with the following transformation:

$$U_i = \frac{X_i - \mu_i}{\sigma_i}. \qquad (7.8)$$

Then, the limit-state function in (7.7) becomes

$$g(\mathbf{X}) = \hat{g}(\mathbf{U}) = a_0 + \sum_{i=1}^{n} a_i(\mu_i + \sigma_i U_i), \qquad (7.9)$$

where $\mathbf{U} = (U_1, \ldots, U_n)$; or

$$\hat{g}(\mathbf{U}) = b_0 + \sum_{i=1}^{n} b_i U_i, \qquad (7.10)$$

where

$$b_0 = a_0 + \sum_{i=1}^{n} a_i \mu_i, \qquad (7.11)$$

and

$$b_i = a_i \sigma_i. \qquad (7.12)$$

Because $Y = \hat{g}(\mathbf{U})$ is a linear combination of normal variables U_i, Y is also normally distributed. Its mean is

$$\mu_Y = b_0, \qquad (7.13)$$

or

$$\mu_Y = a_0 + \sum_{i=1}^{n} a_i \mu_i, \qquad (7.14)$$

and its standard deviation is

$$\sigma_Y = \sqrt{\sum_{i=1}^{n} b_i^2},$$ (7.15)

or

$$\sigma_Y = \sqrt{\sum_{i=1}^{n} (a_i \sigma_i)^2}.$$ (7.16)

The probability of failure p_{f} of the linear limit-state function in (7.7) is then the cumulative distribution function (CDF) $F_Y(0)$; namely,

$$p_{\mathrm{f}} = P\{Y \le 0\} = F_Y(0) = \Phi\left(-\frac{\mu_Y}{\sigma_Y}\right),$$ (7.17)

where Φ is the CDF of a standard normal variable.

Substituting (7.13) and (7.15) into (7.17) yields

$$p_{\mathrm{f}} = \Phi\left(-\frac{b_0}{\sqrt{\sum_{i=1}^{n} b_i^2}}\right).$$ (7.18)

Equation (7.18) provides an analytical solution to the probability of failure for a linear limit-state function with normally distributed random variables. Next, we explain the geometric meaning of $\dfrac{b_0}{\sqrt{\sum_{i=1}^{n} b_i^2}}$ in (7.18).

The distance between a point $\mathbf{U}(U_1, \ldots, U_n)$ in the U-space and the limit-state function $\hat{g}(\mathbf{U}) = 0$ (a hyperplane) is

$$d = \frac{b_0 + \sum_{i=1}^{n} b_i U_i}{\sqrt{\sum_{i=1}^{n} b_i^2}}.$$ (7.19)

Then, the distance between the original $(0, \ldots, 0)$ and $\hat{g}(\mathbf{U}) = 0$ is

$$\beta = \frac{b_0}{\sqrt{\sum_{i=1}^{n} b_i^2}}.$$ (7.20)

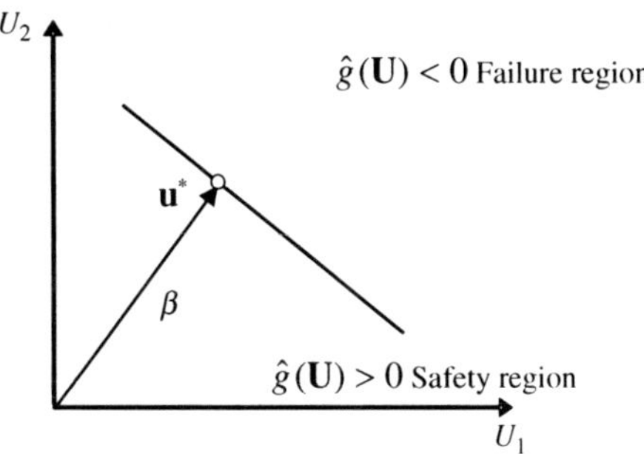

Fig. 7.2 Reliability index and MPP for a linear limit-state function

Based on (7.20), the probability of failure in (7.18) can be simplified as

$$p_\mathrm{f} = \Phi(-\beta). \tag{7.21}$$

β is called the reliability index. Its geometrical meaning is illustrated with a two-dimensional limit-state function in Fig. 7.2. The intersection of the β-line and $\hat{g}(\mathbf{U}) = 0$ is $\mathbf{u}^*$, which is the shortest distance point from the origin to $\hat{g}(\mathbf{U}) = 0$. $\mathbf{u}^*$ is commonly called the MPP because it happens to have the highest probability density on $\hat{g}(\mathbf{U}) = 0$. The reason is explained next.

The marginal PDF of U_i is

$$\phi_{U_i}(u_i) = \frac{1}{\sqrt{2\pi}} \exp\left(-\frac{u_i^2}{2}\right). \tag{7.22}$$

Then, the joint PDF of $\mathbf{U}$ is

$$\phi_{\mathbf{U}}(\mathbf{u}) = \prod_{i=1}^{n} \phi_{U_i}(u_i) = \left(\frac{1}{\sqrt{2\pi}}\right)^n \prod_{i=1}^{n} \exp\left(-\frac{u_i^2}{2}\right)$$

$$= \frac{1}{\left(\sqrt{2\pi}\right)^n} \exp\left(-\frac{\sum_{i=1}^{n} u_i^2}{2}\right) = \frac{1}{\left(\sqrt{2\pi}\right)^n} \exp\left(-\frac{d^2}{2}\right). \tag{7.23}$$

Equation (7.23) indicates that the shorter the distance d, the higher the probability density, where $d = \sqrt{\sum_{i=1}^{n} u_i^2}$ is the distance between point $\mathbf{u}$ and the origin. Because the MPP $\mathbf{u}^*$ is the point closest to the origin, it has the highest probability density on $\hat{g}(\mathbf{U}) = 0$.

Example. X_1 and X_2 are independently and normally distributed. $X_1 \sim N(10, 2^2)$ and $X_2 \sim N(5, 1.5^2)$. The limit-state function is $g(X_1, X_2) = 3X_1 - 2X_2 + 2$. Determine the probability of failure p_f.

Solution According to (7.8), the standard normal variables are $U_1 = \frac{X_1 - \mu_1}{\sigma_1}$ and $U_2 = \frac{X_2 - \mu_2}{\sigma_2}$.

Therefore,

$$X_1 = \mu_1 + U_1\sigma_1 = 10 + 2U_1$$

and

$$X_2 = \mu_2 + U_2\sigma_2 = 5 + 1.5U_2.$$

The limit-state function becomes

$$\hat{g}(U_1, U_2) = 3(10 + 2U_1) - 2(5 + 1.5U_2) + 2 = 22 + 6U_1 - 3U_2.$$

Therefore, $b_0 = 22$, $b_1 = 6$, and $b_2 = -3$.

From (7.20), the reliability index is

$$\beta = \frac{b_0}{\sqrt{b_1^2 + b_2^2}} = \frac{22}{\sqrt{6^2 + (-3)^2}} = 3.2796.$$

According to (7.21),

$$p_f = \Phi(-\beta) = \Phi(-3.2796) = 5.1977 \times 10^{-4}.$$

7.3 First Order Reliability Method

We now discuss how to extend the above approach to nonlinear limit-state functions with nonnormal variables. To use the result obtained above, we need to first transform the general random variables into standard normal variables, and then linearize a nonlinear limit-state function into a linear form like (7.10).

The procedure of the FORM therefore involves two steps: transformation and linearization. As we see, the limit-state function is linearized at the MPP. This way, the minimum accuracy loss may be ensured. The two steps are discussed below.

Step 1: Transformation

As discussed above, $X_i (i = 1, \ldots, n)$ are assumed independent. They are transformed into standard normal variables U_i by

$$F_{X_i}(X_i) = \Phi(U_i) \tag{7.24}$$

where $F_{X_i}(X_i)$ is the CDF of X_i.

For example, if $X_i \sim N(\mu_i, \sigma_i^2)$, then

$$F_{X_i}(X_i) = \Phi\left(\frac{X_i - \mu_i}{\sigma_i}\right),$$

$$U_i = \frac{X_i - \mu_i}{\sigma_i}$$

or

$$X_i = \mu_i + \sigma_i U_i.$$

After the transformation, the limit-state function becomes

$$Y = g(X_1, X_2, \ldots, X_n) = g(F_{X_1}^{-1}[\Phi(U_1)], \ldots, F_{X_n}^{-1}[\Phi(U_n)]) = \hat{g}(\mathbf{U}). \tag{7.25}$$

Step 2: Linearization

$\hat{g}(\mathbf{U})$ is now a function of standard normal random variables. In general, $\hat{g}(\mathbf{U})$ is nonlinear. Even though the original function $g(\mathbf{X})$ is linear, $\hat{g}(\mathbf{U})$ may also be nonlinear because the X-to-U transformation is nonlinear. To use the result obtained for the linear limit-state functions in Sect. 7.2, we need to linearize $\hat{g}(\mathbf{U})$. This linearization will, of course, introduce an error in estimating the probability of failure. Hence, the critical question is: At what point should $\hat{g}(\mathbf{U})$ be linearized? Recall the MPP $\mathbf{u}^*$ has the highest probability density at the limit-state $\hat{g}(\mathbf{U}) = 0$. Therefore, it is natural to use $\mathbf{u}^*$ as the expansion point. This point has the highest contribution to the probability integral in (7.5). As discussed in Sect. 7.2, $\mathbf{u}^*$ is the shortest distance point from $\hat{g}(\mathbf{U})$ to the origin. It can, therefore, be found by solving

$$\begin{cases} \min_{\mathbf{u}} \beta = \|\mathbf{u}\| \\ \text{subject to } \hat{g}(\mathbf{u}) = 0, \end{cases} \tag{7.26}$$

where $\|\cdot\|$ denotes the magnitude of a vector.

With the first-order Taylor series around $\mathbf{u}^*$, $\hat{g}(\mathbf{U})$ is approximated by

$$\hat{g}(\mathbf{U}) \approx \hat{g}(\mathbf{u}^*) + \sum_{i=1}^{n} \left.\frac{\partial \hat{g}}{\partial U_i}\right|_{\mathbf{u}^*} (U_i - u_i^*), \tag{7.27}$$

where u_i^* is the ith component of $\mathbf{u}^*$.

Because $\hat{g}(\mathbf{u}^*) = 0$, $\hat{g}(\mathbf{U})$ can be rewritten as

$$\hat{g}(\mathbf{U}) \approx -\sum_{i=1}^{n} \left.\frac{\partial \hat{g}}{\partial U_i}\right|_{\mathbf{u}^*} u_i^* + \sum_{i=1}^{n} \left.\frac{\partial \hat{g}}{\partial U_i}\right|_{\mathbf{u}^*} U_i = b_0 + \sum_{i=1}^{n} b_i U_i, \tag{7.28}$$

where

$$b_0 = -\sum_{i=1}^{n} \frac{\partial \hat{g}}{\partial U_i}\bigg|_{\mathbf{u}^*} u_i^* \text{ and } b_i = \frac{\partial \hat{g}}{\partial U_i}\bigg|_{\mathbf{u}^*}$$

Since $\hat{g}(\mathbf{U})$ in (7.28) is now linear, according to (7.20), the reliability index is

$$\beta = \frac{b_0}{\sqrt{\sum_{i=1}^{n} b_i^2}} = \frac{-\sum_{i=1}^{n} \frac{\partial \hat{g}}{\partial U_i}\big|_{\mathbf{u}^*} u_i^*}{\sqrt{\sum_{i=1}^{n} \left(\frac{\partial \hat{g}}{\partial U_i}\big|_{\mathbf{u}^*}\right)^2}}. \tag{7.29}$$

As shown in Fig. 7.2, $\mathbf{u}^*$ is perpendicular to the hyperplane $\hat{g}(\mathbf{U}) = 0$ because it is the shortest distance point to $\hat{g}(\mathbf{U}) = 0$. Suppose the gradient of $\hat{g}(\mathbf{U})$ at $\mathbf{u}^*$ is $\nabla \hat{g}(\mathbf{U}) = \left(\frac{\partial \hat{g}}{\partial U_1}, \ldots, \frac{\partial \hat{g}}{\partial U_n}\right)\big|_{u^*}$. As a gradient, $\nabla \hat{g}(\mathbf{U})$ is perpendicular to $\hat{g}(\mathbf{U}) = 0$ and is also in the direction of the steepest increase in $\hat{g}(\mathbf{U})$. Therefore, $\mathbf{u}^*$ is in the opposite direction of $\nabla \hat{g}(\mathbf{U})$. Using their unit vectors, we have

$$\frac{\mathbf{u}^*}{\beta} = -\frac{\nabla \hat{g}(\mathbf{U})}{\|\nabla \hat{g}(\mathbf{U})\|}, \tag{7.30}$$

or

$$\frac{u_i^*}{\beta} = -\frac{\frac{\partial \hat{g}}{\partial U_i}\big|_{\mathbf{u}^*} u_i^*}{\sqrt{\sum_{i=1}^{n} \left(\frac{\partial \hat{g}}{\partial U_i}\big|_{\mathbf{u}^*}\right)^2}} \, (i = 1, \ldots, n). \tag{7.31}$$

Plugging (7.31) into (7.29), we obtain

$$\beta = \|\mathbf{u}^*\| = \sqrt{\sum_{i=1}^{n} (u_i^*)^2}. \tag{7.32}$$

With (7.21), the probability of failure is then computed by

$$p_f = \Phi(-\beta). \tag{7.33}$$

The key of the FORM introduced above is to find the MPP $\mathbf{u}^*$. Once it is found, the reliability index β is simply the distance between $\mathbf{u}^*$ and the origin as shown in (7.32). Then, p_f is computed with (7.33). Next, we discuss how to search for the MPP $\mathbf{u}^*$.

As shown in (7.26), the MPP search is an optimization problem, where the objective is to find the MPP on $\hat{g}(\mathbf{u}) = 0$ to minimize the distance β as shown in

(7.26). Many optimization algorithms, therefore, are applicable for solving (7.26). For higher computational efficiency, however, specialized MPP search algorithms are more commonly used. The most popular HL-RF algorithm [14] is discussed herein.

Equation (7.29) can be rewritten as

$$\beta = \frac{b_0}{\sqrt{\sum\limits_{i=1}^{n} b_i^2}} = \frac{-\sum\limits_{i=1}^{n} \left.\frac{\partial \hat{g}}{\partial U_i}\right|_{\mathbf{u}^*} u_i^*}{\sqrt{\sum\limits_{i=1}^{n} \left(\left.\frac{\partial \hat{g}}{\partial U_i}\right|_{\mathbf{u}^*}\right)^2}} = \frac{-\nabla \hat{g}(\mathbf{u}^*)(\mathbf{u}^*)^{\mathrm{T}}}{\|\nabla \hat{g}(\mathbf{u}^*)\|}. \tag{7.34}$$

We then define a unit vector α along the gradient direction as

$$\alpha(\mathbf{u}^*) = \frac{\nabla \hat{g}(\mathbf{u}^*)}{\|\nabla \hat{g}(\mathbf{u}^*)\|}. \tag{7.35}$$

Then, according to (7.30),

$$\mathbf{u}^* = -\beta \alpha(\mathbf{u}^*). \tag{7.36}$$

The above equation is used in the iterative MPP search process. Suppose at the jth iteration, the U-point is $\mathbf{u}_j$. We linearize $\hat{g}(\mathbf{U})$ at $\mathbf{u}_j$:

$$\hat{g}(\mathbf{U}) = \hat{g}(\mathbf{u}_j) + \nabla \hat{g}(\mathbf{u}_j)(\mathbf{U} - u_j)^{\mathrm{T}}. \tag{7.37}$$

We then find $\mathbf{u}_{j+1}$. Because $\mathbf{u}_{j+1}$ is the MPP of $\hat{g}(\mathbf{U})$, according to (7.36), $\mathbf{u}_{j+1} = -\beta_{j+1}\alpha(\mathbf{u}_j)$. If the algorithm is convergent when $j \to \infty$, according to (7.36), we also have $\mathbf{u}_j = -\beta_j\alpha(\mathbf{u}_j)$. We therefore obtain the equation

$$\begin{aligned} \hat{g}(\mathbf{u}_{j+1}) &= \hat{g}(\mathbf{u}_j) + \nabla \hat{g}(\mathbf{u}_j)(\mathbf{u}_{j+1} - \mathbf{u}_j)^{T} \\ &= \hat{g}(\mathbf{u}_j) - \nabla \hat{g}(\mathbf{u}_j)\alpha(\mathbf{u}_j)^{T}(\beta_{j+1} - \beta_j) = 0. \end{aligned} \tag{7.38}$$

Then, the updated reliability index is

$$\beta_{j+1} = \beta_j + \frac{\hat{g}(\mathbf{u}_j)}{\|\nabla \hat{g}(\mathbf{u}_j)\|}. \tag{7.39}$$

Therefore, the MPP search algorithm can be summarized as

$$\begin{cases} \beta_{j+1} = \beta_j + \dfrac{\hat{g}(\mathbf{u}_j)}{\|\nabla \hat{g}(\mathbf{u}_j)\|} \\ \mathbf{u}_{j+1} = -\beta_{j+1}\alpha(\mathbf{u}_j) \end{cases} \tag{7.40}$$

The termination criteria of the MPP search are as follows: $|\hat{g}_j| < \varepsilon_1$, $|\beta_{j+1} - \beta_j| < \varepsilon_2$, and $\|\mathbf{u}_{j+1} - \mathbf{u}_j\| < \varepsilon_3$, where ε_1, ε_2, and ε_3 are small positive numbers. It is unnecessary to use all of them. One may use two criteria.

As shown above, the MPP search needs the first derivatives of the limit-state function. In many engineering applications, analytical derivatives may not be available, and the numerical methods should be used. If a two-point estimation, such as the forward or backward finite-difference formulae, is used, the estimation of the gradient $\nabla \hat{g}(\mathbf{U})$ will call the limit-state function $\hat{g}(\mathbf{U})$ $n + 1$ times. If we use the number of function calls to measure efficiency, the efficiency of the FORM will be equal to $N(n + 1)$, where N is the number of iterations of the MPP search. Hence, the efficiency of the FORM is directly proportional to the dimensionality of the limit-state function.

The HR-LF algorithm is efficient for the MPP search, but may not converge for highly nonlinear limit-state functions. Zhang and Der Kiureghian propose an improved HL-RF algorithm, denoted by i-HRLF [23]. It is globally convergent because it can converge to a local MPP from any starting point. Other MPP search algorithms have also been reported in literature, including those in [1, 9].

Example. The voltage divider with two resistors [15] is shown in Fig. 7.3. The two resistor values R_1 and R_2 are normal random variables. The input voltage V_{in} follows a lognormal distribution. The information of the random variables is summarized in Table 7.1. The design requirement is to have an open circuit output voltage V_{out} larger than 7 V. Then, the limit-state function is defined as

$$g(X) = \frac{R_2}{(R_1 + R_2)} V_{\text{in}} - 7$$

or

$$g(X) = R_2 V_{\text{in}} - 7(R_1 + R_2).$$

Solution. 1 Transformation from $\mathbf{X}$ to $\mathbf{U}$

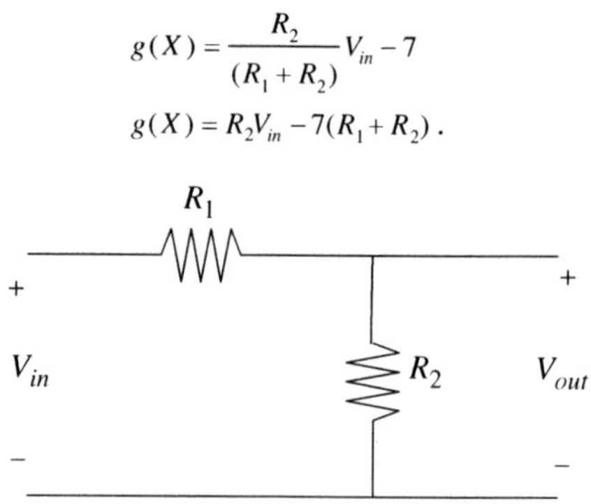

Fig. 7.3 Voltage divider

Table 7.1 Random variables

Random variable	Mean value	Standard deviation	Distribution type
R_1 (Ω)	100	4	Normal
R_2 (Ω)	300	6	Normal
V_{in} (V)	10	0.1	Lognormal

According to (7.8), for R_1 and R_2,

$$U_1 = \frac{R_1 - \mu_{R_1}}{\sigma_{R_1}}$$

and

$$U_2 = \frac{R_2 - \mu_{R_2}}{\sigma_{R_2}}.$$

Then,

$$R_1 = \mu_{R_1} + U_1 \sigma_{R_1} = 100 + 4U_1,$$

and

$$R_2 = \mu_{R_2} + U_2 \sigma_{R_2} = 300 + 6U_2.$$

Because V_{in} is lognormally distributed, $\ln V_{\text{in}} \sim N(\mu'_{V_{\text{in}}}, (\sigma'_{V_{\text{in}}})^2)$, where

$$\sigma'_{V_{\text{in}}} = \left\{ \ln\left[1 + \left(\frac{\sigma_{V_{\text{in}}}}{\mu_{V_{\text{in}}}}\right)^2\right] \right\}^{\frac{1}{2}} = \left\{ \ln\left[1 + \left(\frac{0.1}{10}\right)^2\right] \right\}^{\frac{1}{2}} = 0.01$$

$$\mu'_{V_{\text{in}}} = \ln(\mu_{V_{\text{in}}}) - \tfrac{1}{2}\sigma'^2_{V_{\text{in}}} = \ln(10) - \tfrac{1}{2} \times 0.01^2 = 2.3025.$$

Thus, $\ln V_{\text{in}} \sim N(2.3025, 0.01^2)$.
According to (7.8),

$$U_3 = \frac{\ln V_{\text{in}} - \mu'_{V_{\text{in}}}}{\sigma'_{V_{\text{in}}}}$$

or

$$V_{\text{in}} = \exp(\mu'_{V_{\text{in}}} + U_3 \sigma'_{V_{\text{in}}}) = \exp(2.3025 + 0.01U_3)$$

Therefore, the limit-state function after the transformation becomes

$$\hat{g}(\mathbf{X}) = (300 + 6U_2)\exp(2.3025 + 0.01U_3) - 7(400 + 4U_1 + 6U_2).$$

2. MPP search

The gradient $\nabla \hat{g}(\mathbf{U}) = \left(\frac{\partial \hat{g}}{\partial U_1}, \frac{\partial \hat{g}}{\partial U_2}, \frac{\partial \hat{g}}{\partial U_3} \right)$ required for the MPP search is first evaluated as follows:

$$\frac{\partial \hat{g}}{\partial U_1} = -28,$$

$$\frac{\partial \hat{g}}{\partial U_2} = 6 \exp(2.3025 + 0.01 U_3) - 42,$$

$$\text{and } \frac{\partial \hat{g}}{\partial U_3} = (3 + 0.06 U_2) \exp(2.3025 + 0.01 U_3).$$

The convergence criteria are $|\hat{g}_k| < \varepsilon_1$ and $\|\mathbf{u}_k - \mathbf{u}_{k-1}\| < \varepsilon_2$, where $\varepsilon_1 = \varepsilon_2 = 0.001$.

Then, we start the MPP search from the origin $\mathbf{u}_0 = (0, 0, 0)$.

Iteration 1: $k = 1$

The limit-state function $\mathbf{u}_0$ is

$$\hat{g}(\mathbf{u}_0) = 300 \exp(2.3025) - 7(400) = 199.7447.$$

Its gradient at $\mathbf{u}_0$ is computed by

$$\nabla \hat{g}(\mathbf{u}_0) = (-28, 6 \exp(2.3025) - 42, 3 \exp(2.3025))$$
$$= (-28, 17.9949, 29.9974).$$

Therefore,

$$\alpha_0 = \frac{\nabla \hat{g}(\mathbf{u}_0)}{\|\nabla \hat{g}(\mathbf{u}_0)\|} = \frac{(-28, 17.9949, 29.9974)}{\sqrt{-28^2 + 17.9949^2 + 29.9974^2}}$$
$$= (-0.6249, 0.40161, 0.66948).$$

The reliability index at the origin is

$$\beta_0 = 0,$$

and the updated reliability index is given by (7.40) as

$$\beta_1 = \beta_0 + \frac{\hat{g}(\mathbf{u}_0)}{\|\nabla \hat{g}(\mathbf{u}_0)\|} = \frac{199.7447}{\sqrt{-28^2 + 17.9949^2 + 29.9974^2}} = 4.4579.$$

Then, the updated MPP is

$$\mathbf{u}_1 = -\beta_1 \alpha_0 = -4.4579(-0.6249, 0.40161, 0.66948)$$
$$= (2.7858, -1.7903, -2.9845).$$

Iteration 2: $k = 2$

$$\hat{g}(\mathbf{u}_1) = [300 + 6(-1.7903)] \exp[2.3025 + 0.01(-2.9845)]$$
$$- 7[400 + 4(2.7858) + 6(-2.9845)]$$
$$= 4.4811$$

$$\frac{\partial \hat{g}(\mathbf{u}_1)}{\partial U_1} = -28,$$

$$\frac{\partial \hat{g}(\mathbf{u}_1)}{\partial U_2} = 6 \exp(2.3025 + 0.01 U_3) - 42$$

$$= 6 \exp[2.3025 + 0.01(-2.9845)] = 16.2308$$

$$\frac{\partial \hat{g}(\mathbf{u}_1)}{\partial U_3} = (3 + 0.06 U_2) \exp(2.3025 + 0.01 U_3)$$

$$= [3 + 0.06(-1.7903)] \exp[2.3025 + 0.01(-2.9845)]$$
$$= 28.0729$$

Thus,

$$\nabla \hat{g}(\mathbf{u}_1) = (-28, 16.2308, 28.0729)$$

$$\alpha_1 = \frac{\nabla \hat{g}(\mathbf{u}_1)}{\|\nabla \hat{g}(\mathbf{u}_1)\|} = \frac{(-28, 17.9949, 29.9974)}{\sqrt{-28^2 + 16.2308^2 + 28.0729^2}}$$
$$= (-0.6249, 0.40161, 0.66948).$$

The reliability index is updated by

$$\beta_2 = \beta_1 + \frac{\hat{g}(\mathbf{u}_1)}{\|\nabla \hat{g}(\mathbf{u}_1)\|} = 4.4579 + \frac{4.4811}{\sqrt{-28^2 + 16.2308^2 + 28.0729^2}}$$
$$= 4.5625,$$

and the MPP is updated by

$$\mathbf{u}_2 = -\beta_2 \alpha_1 = -4.5625(-0.6249, 0.40161, 0.66948)$$
$$= (2.9818, -1.7285, -2.9896).$$

Because

$$|\hat{g}(\mathbf{u}_1)| = 4.4811 > \varepsilon_1 = 0.001$$

and

$$\|\mathbf{u}_2 - \mathbf{u}_1\| = \|(2.9818, -1.7285, -2.9896) - (2.7858, -1.7903, -2.9845)\|$$
$$= 0.2056 > \varepsilon_2 = 0.001,$$

we continue the search and repeat the process in iteration 3. The details of iteration 3 are omitted. We proceed to iteration 4.

Iteration 4: $k = 4$

$$\mathbf{u}_3 = (2.9781, -1.726, -2.9895),$$

$$\hat{g}(\mathbf{u}_3) = -3.9341 \times 10^{-5}$$

$$\nabla \hat{g}(\mathbf{u}_3) = (-28, 16.2279, 28.1089)$$

Thus,

$$\alpha_3 = (-0.6532, 0.37858, 0.65575),$$

$$\beta_4 = \beta_3 + \frac{\hat{g}(\mathbf{u}_3)}{\|\nabla \hat{g}(\mathbf{u}_3)\|} = 4.5591 + \frac{-3.9341 \times 10^{-5}}{\sqrt{-28^2 + 16.2279^2 + 28.1089^2}} = 4.559,$$

where β_3 is obtained from iteration 3.

Then, the MPP is updated as

$$\mathbf{u}_4 = -\beta_4 \alpha_3 = -4.559(-0.6532, 0.37858, 0.65575)$$
$$= (2.978, -1.7259, -2.9896).$$

Since

$$|\hat{g}(\mathbf{u}_4)| = -6.4494 \times 10^{-8} < \varepsilon_1 = 0.001$$

and

$$\|\mathbf{u}_4 - \mathbf{u}_3\| = 1.732 \times 10^{-4} < \varepsilon_2 = 0.001,$$

both of the convergence criteria are satisfied. We then stop the search.

Therefore, the MPP is found at $\mathbf{u}^* = \mathbf{u}_4 = (2.978, -1.7259, -2.9896)$, and the reliability index is $\beta = \beta_4 = 4.559$.

3. p_f estimation

From (7.21),

$$p_f^{\text{FORM}} = \Phi(-\beta) = \Phi(-4.559) = 2.5693 \times 10^{-6}.$$

To access the accuracy of the FORM for this problem, reliability assessment based on Monte Carlo simulation (MCS) is used as a reference. To ensure that the MCS approach produces an accurate estimation, we use 10^9 sample points. The point estimation is

$$p_f^{\text{MCS}} = 2.627 \times 10^{-6}.$$

It is seen that the FORM solution is very close to the MCS solution. The relative error of the FORM is

$$\varepsilon^{\text{FORM}} = \frac{\left|p_f^{\text{FORM}} - p_f^{\text{MCS}}\right|}{p_f^{\text{MCS}}} 100\% = \frac{\left|2.5693 \times 10^{-6} - 2.627 \times 10^{-6}\right|}{2.627 \times 10^{-6}} 100\%$$
$$= 2.20\%.$$

Note that the 95% confidence interval of the MCS solution is $[2.527 \times 10^{-6}, 2.727 \times 10^{-6}]$ given the sample size of 10^9.

7.4 Second-Order Reliability Method

The FORM uses the first-order approximation to the limit-state function. If the limit-state function in the U-space is highly nonlinear and the variance range of uncertain variables is large, the error of the reliability estimation will be large. If higher accuracy is desired, we can approximate the limit-state function with the second order Taylor expansion series at the MPP. The associated method is then referred to as the SORM [2]. In general, the SORM is more accurate than the FORM but is less efficient because the second derivatives are required.

As in the FORM, random variables $\mathbf{X}$ are first transformed into standard normal variables $\mathbf{U}$. Then, the second-order approximation $\hat{g}(\mathbf{U})$ is derived as

$$\hat{g}(\mathbf{U}) \approx \hat{g}(\mathbf{u}^*) + \sum_{i=1}^{n} \left.\frac{\partial \hat{g}}{\partial U_i}\right|_{\mathbf{u}^*} (U_i - u_i^*) + \frac{1}{2} \sum_{i=1}^{n} \sum_{j=1}^{n} \left.\frac{\partial^2 \hat{g}}{\partial U_i \partial U_j}\right|_{\mathbf{u}^*} (U_i - u_i^*)(U_j - u_j^*).$$

$$(7.41)$$

Writing (7.41) into a vector form and noticing that $\hat{g}(\mathbf{u}^*) = 0$, we have

$$\hat{g}(\mathbf{U}) \approx \nabla g(\mathbf{u}^*)(\mathbf{U} - \mathbf{u}^*)^T + \frac{1}{2}(\mathbf{U} - \mathbf{u}^*)\nabla^2 g(\mathbf{u}^*)(\mathbf{U} - \mathbf{u}^*)^T, \qquad (7.42)$$

where $\nabla^2 g$ is the Hessian matrix, the symmetric square matrix of second partial derivatives of $\hat{g}(\mathbf{U})$.

Analytical solution to the probability of failure cannot be directly derived from (7.42) because of the second order terms. Linear and orthogonal transformations are needed to simplify (7.41) or (7.42) further. After the transformations, $\hat{g}(\mathbf{U})$ will be a function of independent and standard normal variables. Once the transformation is complete, according to [2], the probability of failure is estimated by

$$p_{\mathrm{f}} = \Phi(-\beta) \prod_{i=1}^{n-1} (1 + \beta k_i)^{-\frac{1}{2}}, \qquad (7.43)$$

where β is the reliability index, and $k_i (i = 1, \ldots, n-1)$ are the principal curvatures of $\hat{g}(\mathbf{U})$ at the MPP. Comparing the above approximation with the FORM approximation in (7.31), we see that $\prod_{i=1}^{n-1} (1 + \beta k_i)^{-\frac{1}{2}}$ is a correction term for the FORM solution.

Next, we discuss how to compute the principal curvatures k_i.

k_i are computed as the eigenvalues of a matrix $\mathbf{A}$, whose elements are given by

$$a_{ij} = \frac{\left(\mathbf{R}\nabla^2 \hat{g}(\mathbf{u}^*)\mathbf{R}^{\mathrm{T}}\right)_{ij}}{\|\nabla \hat{g}(\mathbf{u}^*)\|} \quad i,j = 1, \ldots, n-1, \qquad (7.44)$$

where $\mathbf{R}$ is an $n \times n$ rotation matrix, which is obtained from another matrix $\mathbf{R}_0$ with the Gram–Schmidt orthogonalization. $\mathbf{R}_0$ is given by

$$\mathbf{R}_0 = \begin{bmatrix} \mathbf{I}_{n-1} & 0 \\ -\alpha_1 - \alpha_2 \cdots - \alpha_{n-1} & -\alpha_n \end{bmatrix}, \qquad (7.45)$$

where $\mathbf{I}_{n-1}$ is an $(n-1) \times (n-1)$ identity matrix, and the last row of $\mathbf{R}_0$ is the unit vector $-\alpha$ defined in (7.35).

The general procedure of the SORM is as follows.

1. Search the MPP

This step is the same as the MPP search in the FORM mentioned above. After $\mathbf{u}^*$ is found, compute β, $\alpha(\mathbf{u}^*)$, $\nabla(\mathbf{u}^*)$, and $\nabla^2(\mathbf{u}^*)$.

2. Construct matrix $\mathbf{R}$ with Gram–Schmidt orthogonalization

Let the rows of $\mathbf{R}_0$ and $\mathbf{R}$ be $\mathbf{r}_{0i}$ and $\mathbf{r}_i (i = 1, \ldots, n)$, respectively. $\mathbf{R}$ can be generated from $\mathbf{R}_0$ with the following Gram–Schmidt orthogonalization procedure:

$$\begin{cases} \mathbf{r}_n = \mathbf{r}_{0n} \\ \mathbf{r}'_k = \mathbf{r}_{0k} - \sum_{j=k+1}^{n} \frac{\mathbf{r}_j \mathbf{r}_{0k}^T}{\mathbf{r}_j \mathbf{r}_j^T} \mathbf{r}_j, \mathbf{r}_k = \frac{\mathbf{r}'_k}{\|\mathbf{r}'_k\|}, & k = n-1, n-2, \ldots, 1. \end{cases} \qquad (7.46)$$

3. Compute matrix $\mathbf{A}$

   Compute matrix $\mathbf{A}$ using (7.44). Note that the size of $\mathbf{A}$ is $(n-1) \times (n-1)$.

4. Calculate the principal curvatures $k_i(i = 1, \ldots, n-1)$.

   Solving the eigenvalue problem

   $$\mathbf{A}\mathbf{Z} = k\mathbf{Z}. \tag{7.47}$$

where $\mathbf{Z}$ is a vector with a length of $n-1$. We then solve the equation as follows:

$$\det(\mathbf{A} - kI) = 0, \tag{7.48}$$

where det denotes the determinant of a matrix.

Then, the eigenvalues are obtained

$$\mathbf{k} = (k_1, \ldots, k_{n-1})^{\mathrm{T}}. \tag{7.49}$$

5. Compute the probability of failure p_f with (7.43).

   The comparison between the SORM and FORM is shown in Fig. 7.4. As indicated in the figure, the SORM is in general more accurate than the FORM for a nonlinear limit-state function $\hat{g}(\mathbf{U})$. The SORM is derived based on asymptotic approximation. In other words, it is accurate when β is sufficiently large. It may not be accurate when β is small. Several improvements have been made to remedy the problem as described in [20]. Other forms of the SORM formulae have also proposed. Some of them are reported in [3, 16, 24].

Example. Here, we use the SORM to solve the same example presented in Sect. 7.3.

Solution

1. Search for the MPP
   In the problem presented above, we have obtained the MPP $\mathbf{u}^* = (2.978, -1.7259, -2.9896)$, the gradient of the limit-state function at the MPP $\nabla\hat{g}(\mathbf{u}^*) = (-28, 16.2278, 28.1089)$, the unit vector $\alpha(\mathbf{u}^*) = (-0.65321, 0.37858, 0.65575)$, and the reliability index $\beta = 4.559$.

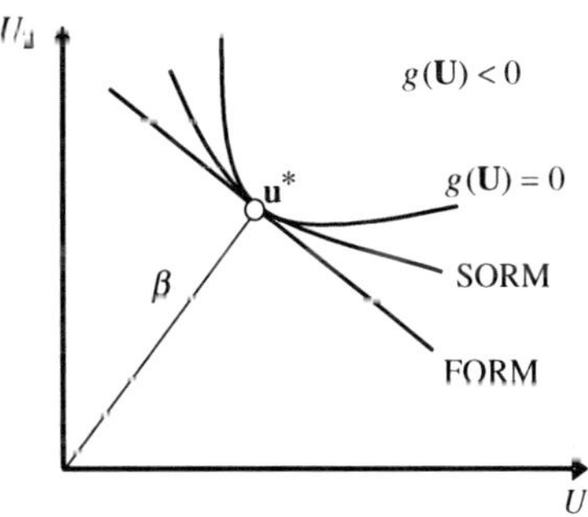

Fig. 7.4 FORM vs. SORM

The Hessian matrix $\nabla^2(\mathbf{u}^*)$ requires second derivatives, which are given by

$$\frac{\partial^2 \hat{g}}{\partial U_1^2} = 0, \quad \frac{\partial^2 \hat{g}}{\partial U_1 \partial U_2} = 0, \quad \frac{\partial^2 \hat{g}}{\partial U_1 \partial U_3} = 0,$$

$$\frac{\partial^2 \hat{g}}{\partial U_2^2} = 0, \quad \frac{\partial^2 \hat{g}}{\partial U_2 \partial U_3} = 0.06 \exp(2.3025 + 0.01 U_3),$$

and $\dfrac{\partial^2 \hat{g}}{\partial U_3^2} = (0.03 + 6 \times 10^{-4} U_2) \exp(2.3025 + 0.01 U_3).$

Thus,

$$\nabla^2(\mathbf{u}^*) = \left. \begin{bmatrix} \frac{\partial^2 \hat{g}}{\partial U_1^2} & \frac{\partial^2 \hat{g}}{\partial U_1 \partial U_2} & \frac{\partial^2 \hat{g}}{\partial U_1 \partial U_3} \\ \frac{\partial^2 \hat{g}}{\partial U_2 \partial U_1} & \frac{\partial^2 \hat{g}}{\partial U_2^2} & \frac{\partial^2 \hat{g}}{\partial U_2 \partial U_3} \\ \frac{\partial^2 \hat{g}}{\partial U_3 \partial U_1} & \frac{\partial^2 \hat{g}}{\partial U_3 \partial U_2} & \frac{\partial^2 \hat{g}}{\partial U_3^2} \end{bmatrix} \right|_{U=\mathbf{u}^*}$$

$$= \begin{bmatrix} 0 & 0 & 0 \\ 0 & 0 & 0.58228 \\ 0 & 0.58228 & 0.28109 \end{bmatrix}.$$

2. Construct matrix $\mathbf{R}$

First, we construct matrix $\mathbf{R}_0$ with (7.44).

$$\mathbf{R}_0 = \begin{bmatrix} 1 & 0 & 0 \\ 0 & 1 & 0 \\ -\alpha_1 & -\alpha_2 & -\alpha_3 \end{bmatrix} = \begin{bmatrix} 1 & 0 & 0 \\ 0 & 1 & 0 \\ 0.65321 & -0.37858 & -0.65575 \end{bmatrix}.$$

We then use (7.45) to compute matrix $\mathbf{R}$ with $\mathbf{R}_0$.

$$\mathbf{r}_3 = \mathbf{r}_{03} = (0.65321, -0.37858, -0.65575)$$

$$\mathbf{r}_2' = \mathbf{r}_{02} - \sum_{j=3}^{3} \frac{\mathbf{r}_j \mathbf{r}_{02}^T}{\mathbf{r}_j \mathbf{r}_j^T} \mathbf{r}_j = \mathbf{r}_{02} - \frac{\mathbf{r}_3 \mathbf{r}_{02}^T}{\mathbf{r}_3 \mathbf{r}_3^T} \mathbf{r}_3$$

$$= (0, 1, 0) - \frac{(0.65321, -0.37858, -0.65575)(0, 1, 0)^T}{\|\alpha(\mathbf{u}^*)\|}$$

$$\cdot (0.65321, -0.37858, -0.65575)$$

$$= (0.24729, 0.85668, -0.24825)$$

$$\mathbf{r}_2 = \frac{\mathbf{r}_2'}{\|\mathbf{r}_2'\|} = (0.26717, 0.92557, -0.26821)$$

$$\mathbf{r}_1' = \mathbf{r}_{01} - \sum_{j=2}^{3} \frac{\mathbf{r}_j \mathbf{r}_{01}^T}{\mathbf{r}_j \mathbf{r}_j^T} \mathbf{r}_j = \mathbf{r}_{01} - \frac{\mathbf{r}_2 \mathbf{r}_{01}^T}{\mathbf{r}_2 \mathbf{r}_2^T} \mathbf{r}_2 - \frac{\mathbf{r}_3 \mathbf{r}_{01}^T}{\mathbf{r}_3 \mathbf{r}_3^T} \mathbf{r}_3 = (0.50194, 0, 0.5000)$$

and

$$\mathbf{r}_1 = \frac{\mathbf{r}_1'}{\|\mathbf{r}_1'\|} = (0.70848, 0, 0.70573).$$

Therefore,

$$\mathbf{R} = \begin{bmatrix} \mathbf{r}_1 \\ \mathbf{r}_2 \\ \mathbf{r}_3 \end{bmatrix} = \begin{bmatrix} 0.70848 & 0 & 0.70573 \\ 0.26717 & 0.92557 & -0.26821 \\ 0.65321 & -0.37858 & -0.65575 \end{bmatrix}.$$

3. Construct matrix A
   According to (7.43),

$$\frac{(\mathbf{R}\nabla^2 \hat{g}(\mathbf{u}^*)\mathbf{R}^T)}{\|\nabla \hat{g}(\mathbf{u}^*)\|} = \begin{bmatrix} 0.00327 & 0.00763 & -0.00666 \\ 0.00763 & -0.00627 & -0.00571 \\ -0.00666 & -0.00571 & 0.00956 \end{bmatrix}.$$

Thus,

$$\mathbf{A} = \begin{bmatrix} 0.00327 & 0.00763 \\ 0.00763 & -0.00627 \end{bmatrix}.$$

4. Calculate the principal curvatures k_1 and k_2.
   Solve the eigenvalue problem

$$\mathbf{A}\mathbf{Z} = k\mathbf{Z}.$$

$$\det(\mathbf{A} - kI) = \det \begin{bmatrix} 0.00327 - k & 0.00763 \\ 0.00763 & -0.00627 - k \end{bmatrix} = 0$$

We then obtain the eigenvalues $k_1 = 0.00750$ and $k_2 = -0.01050$.

5. Compute p_f
   From (7.42), the probability of failure is

$$p_f^{SORM} = \Phi(-\beta) \prod_{i=1}^{n-1} (1 + \beta k_i)^{-\frac{1}{2}}$$

$$= \Phi(-4.559)(1 + 4.559 \times 0.00750)^{-\frac{1}{2}}[1 + 4.559 \times (-0.01050)]^{-\frac{1}{2}}$$

$$= 2.5892 \times 10^{-6}$$

The result of MCS method is also used for accuracy comparison. The error of the SORM is

$$\varepsilon^{SORM} = \frac{\left\| p_f^{SORM} - p_f^{MCS} \right\|}{p_f^{MCS}} 100\% = \frac{\left\| 2.5892 \times 10^{-6} - 2.627 \times 10^{-6} \right\|}{2.627 \times 10^{-6}} 100\%$$
$$= 1.46\%.$$

Because $\varepsilon^{SORM} < \varepsilon^{FORM} = 2.20\%$, the SORM is more accurate than the FORM for this problem.

7.5 Other Topics of the MPP-Based Methods

7.5.1 Dependent Random Variables

In the above methods, we assumed that all the random variables are mutually independent. The methods are also applicable for dependent random variables. This can be achieved by transforming the dependent basic random variables **X** into independently and normally distributed random variables **U**. The transformation is given by [19].

$$\begin{cases} U_1 = \Phi^{-1}\{F_{X_1}(X_1)\} \\ U_2 = \Phi^{-1}\{F_{X_2|X_1}(X_2|X_1)\} \\ U_3 = \Phi^{-1}\{F_{X_3|X_1,X_2}(X_3|X_1,X_2)\}, \\ \vdots \end{cases} \tag{7.50}$$

where F_{X_1} is the CDF of X_1, and $F_{X_2|X_1}$ and $F_{X_3|X_1,X_2}$ are conditional CDFs of X_2 and X_3, respectively. Once the transformation is complete, the same procedure of the FORM or SORM can be followed.

7.5.2 MPP-Based Monte Carlo Simulation

Compared with MCS described in Chap. 4, the MPP-based methods are much more efficient when the probability of failure is low or the reliability is high. This has been shown in the examples in Sects. 7.3 and 7.4. For highly nonlinear limit-state functions, however, the MPP-based methods may not be accurate. This can be remedied by performing MCS around the MPP.

As shown in Fig. 7.5, if the direct MCS is used, there may not be any failure points for a high reliability problem. All the samples (the lower cloud in Fig. 7.5) are in the safety region. To better estimate the probability of failure without increasing

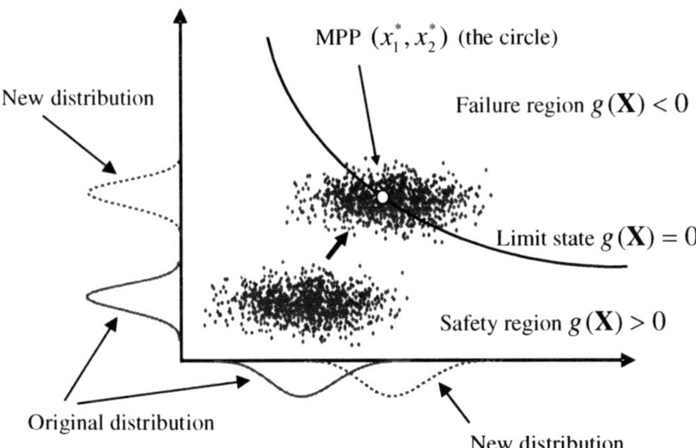

Fig. 7.5 MPP-based importance sampling

the sample size, we shift the mean values of the distributions of the basic random variables to the MPP. As shown in the figure, the entire cloud of the samples is moved toward the failure region, and a significant number of samples are now in the failure region.

The above approach belongs to the importance sampling methods. It generates samples from "biased" distributions (the shifted distributions) rather than the original distributions of $\mathbf{X}$ so that more samples will be in the failure region. Then, the result of the probability of failure is weighted to correct for the use of the biased distributions. This ensures that the new estimator of the probability of failure be unbiased.

7.5.3 MPP-Based Methods in the Original X-Space

The FORM requires the X-to-U transformation. For nonnormal distributions, this transformation is nonlinear. As a result, a linear limit-state function with nonnormal distributions in the X-space will become nonlinear in the U-space. A nonlinear limit-state function with nonnormal distributions in the X-space may be more nonlinear in the U space. Consequently, the X-to-U transformation may result in a larger error when the transformation increases the nonlinearity of the limit-state function. To remedy this problem, Du and Sudjianto have developed the MPP-based method implemented only in the X-space. The method is called the first-order saddlepoint approximation (FOSPA) method [12].

The FOSPA method first searches for the MPP in the X-space. Similar to the FORM, the MPP $\mathbf{x}^*$ is found by maximizing the joint probability density of the basic random variables $\mathbf{X}$ on the limit state $g(\mathbf{X}) = 0$; then, the limit-state function $g(\mathbf{X})$ is linearized at the MPP $\mathbf{x}^*$. Figure 7.6 shows a bivariate limit-state function

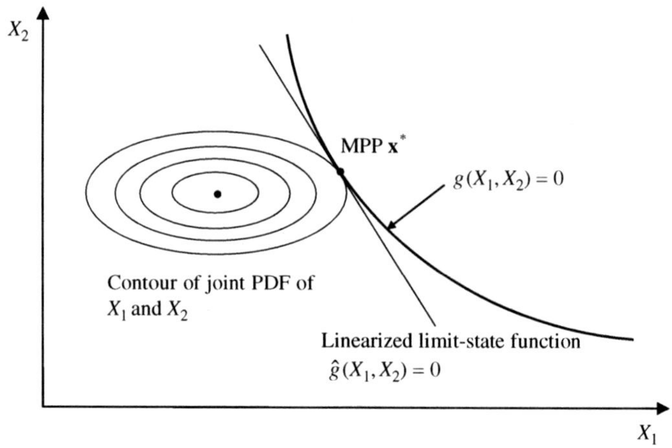

Fig. 7.6 First-order saddlepoint approximation

$g(X_1, X_2) = 0$, the MPP, and the linearized function $\hat{g}(X_1, X_2)$. Based on the linearized limit-state function, saddlepoint approximation [5] is then applied to estimate the probability of failure $p_f \approx \Pr\{\hat{g}(\mathbf{X}) \le 0\}$.

The FOSPA method does not directly evaluate the probability $\Pr\{\hat{g}(\mathbf{X}) \le 0\}$ from the joint PDF of $\mathbf{X}$ because the task is difficult. Instead, it approximates the probability using the cumulant generating function (CGF), $K_{\hat{g}}$, of the linear function $\hat{g}(\mathbf{X})$. $K_{\hat{g}}$ can be readily obtained from the CGFs of the individual random variables in $\mathbf{X}$ because $\hat{g}(\mathbf{X})$ is a linear combination of $\mathbf{X}$. Even though the theory of saddlepoint approximation is complicated, its application is fairly straightforward as shown in [12]. The reader is referred to [12] for detailed procedure of the FOSPA method.

Saddlepoint approximation is very accurate in estimating tail probabilities for a linear function. For the linearized limit-state function $\hat{g}(\mathbf{X})$, the error of the saddlepoint approximation is negligible. Therefore, the FOSPA method is more accurate than the FORM if the latter method increases the nonlinearity of the limit-state function [7]. Since the FOSPA method also requires the MPP search, its efficiency is similar to that of the FORM. It is noted that if all the ransom variables are normally distributed, both the FORM and FOSPA method will produce an identical solution [12].

7.6   Conclusions

Because of their good accuracy and efficiency, MPP-based methods have been widely used in probabilistic engineering analysis, especially in structural reliability analysis. In general, the SORM is more efficient than the FORM, though counter-examples exist. Since the SORM needs the second derivatives, it is not as efficient as the FORM. Both the FORM and SORM can provide satisfactory results for most of engineering problems with high reliability. However, the efficiency of the MPP-based methods decreases with the increase of the number of random variables. Hence, compared with MCS, the advantage of the MPP-based methods may diminish when the reliability is low and the number of random variables is huge.

The FORM is also commonly used in reliability-based design optimization (RBDO) [4] because of the high efficiency of the FORM. RBDO is a design methodology that minimizes the cost of a product and at the same time maintains reliability to a desired (or required) level. During the optimization process, the design point is updated. At each of the new points, the reliability associated with each constraint (requirement) needs to be calculated. This is where the FORM is used. A vast amount of literature about RBDO exists. Examples of the recent development in FORM-based RBDO can be found in [10, 11, 17, 21, 22, 25].

The MPP-based methods have also recently been extended to applications when interval variables exist. Examples of intervals include dimension tolerances, measurement errors, and product lives observed by periodical monitoring. Furthermore, if samples are limited, obtaining the distribution of a random variable may not be possible. In this case, the random variable can also be represented by an interval. When both interval and random variables are present, the reliability will also be an interval. Several MPP-based methods have been developed to calculate the lower and upper bounds of the probability of failure. Du has developed a unified uncertainty analysis method for the probability bound using the FORM [8]. He has also considered both interval and random variables in RBDO [13]. The incorporation of interval variables in RBDO using the MPP have also been reported in [6, 18].

References

1. Awad M, Singh N, Sudjianto A (2006) MPP search method using limited radial importance sampling. Int J Prod Dev 3(3–4):305–318
2. Breitung K (1984) Asymptotic approximations for multinormal integrals. ASCE J Eng Mech 110(3):357–366
3. Cai G, Elishakoff I (1994) Refined second-order reliability analysis. Struct Saf 14:267–276
4. Choi S, Grandhi RV, Canfield RA (2006) Reliability-based structural design. Springer, New York
5. Daniels HE (1954) Saddlepoint approximations in statistics. Ann Math Stat 25:631–650
6. Du L, Choi KK, Youn BD et al (2006) Possibility-based design optimization method for design problems with both statistical and fuzzy input data. J Mech Des 128:928–935
7. Du X (2008) Saddlepoint approximation for sequential optimization and reliability analysis. ASME J Mech Des 30(1):011011–011022
8. Du X (2008) Unified uncertainty analysis with the first order reliability method. ASME J Mech Des 130(9):091401–091410
9. Du X, Chen W (2001) A most probable point based method for uncertainty analysis. J Des and Manuf Autom 4(1):47–66
10. Du X, Chen W (2004) Sequential optimization and reliability assessment for probabilistic design. ASME J Mech Des 126(2):225–233

11. Du X, Huang B (2007) Reliability-based design optimization with equality constraints. Int J Numer Methods in Eng 2(11):1314–1331
12. Du X, Sudjianto A (2004) The first order saddlepoint approximation for reliability analysis. AIAA J 42(6):1199–1207
13. Du X, Sudjianto A, Huang B (2005) Reliability-based design under the mixture of random and interval variables. ASME J Mech Des 127(6):1068–1076
14. Hasofer AM, Lind NC (1974) Exact and invariant second-moment code format. J Eng Mech Div 100(1):111–121
15. Kamas LA, Sanders SR (1994) Reliability analysis via numerical simulation of power electroniccircuits. Comput Power Electron, IEEE 4th Workshop on 7–10 Aug 1994
16. Köylüoglu HU, Nielsen SRK (1994) New approximations for SORM integrals. Struct Saf 13 (4):235–246
17. Liang J, Mourelatos ZP, Tu J (2008) A single-loop method for reliability-based design optimization. Int J Prod Dev 5(1/2):76–92
18. Mourelatos ZP, Zhou J (2006) A design optimization method using evidence theory. J Mech Des 128(4):901–908
19. Rosenblatt M (1952) Remarks on a multivariate transformation. Ann Math Stat 23:470–472
20. Tvedt L (1990) Distribution of quadratic forms in normal space: application to structural reliability. J Eng Mech 116(6):1183–1197
21. Wu YT, Shin Y, Sues R et al (2001) Safety factor based approach for probability-based design optimization. 42nd AIAA/ ASME/ASCE/AHS/ASC Struct, Struct Dyn and Mater Conf, Seattle, WA, 2001, AIAA Paper 2001-1522
22. Youn BD, Choi KK, Park YH (2003) Hybrid analysis method for reliability-based design optimization. J Mech Des 125(2):221–232
23. Zhang Y, Der Kiureghian A (1994) Two improved algorithms for reliability analysis. 6th IFIP WG7.5 Work Conf on Reliab and Optim Struct Syst, Assisi, Italy, Sept 7–9 1994
24. Zhao Y, Ono T (1999) New approximations for SORM. J Eng Mech 125(1):79–93
25. Zou T, Mahadevan S (2006) A direct decoupling approach for efficient reliability-based design optimization. Struct Multidiscip Optim 31(3):190–200

Chapter 8
Extreme Value Theory: Application to Memory Statistics

Robert C. Aitken, Amith Singhee, and Rob A. Rutenbar

8.1 Introduction

Device variability is increasingly important in memory design, and a fundamental question is how much design margin is enough to ensure high quality and robust operation without overconstraining performance. For example, it is very unlikely that the "worst" bit cell is associated with the "worst" sense amplifier, making an absolute "worst-case" margin method overly conservative, but this assertion needs to be formalized and tested. Standard statistical techniques tend to be of limited use for this type of analysis, for two primary reasons: First, worst-case values by definition involve the tails of distributions, where data is limited and normal approximations can be poor. Second, the worst-case function itself does not lend itself to simple computation (the worst case of a sum is not the sum of worst cases, for example). These concepts are elaborated later in this chapter.

Extreme value theory (EVT) has been developed to address these and other points. This chapter shows how EVT can be used to work with limited data and make reasonable predictions about memory margining and other statistical aspects of memory design.

8.1.1 Design Margin and Memory

Memory product designs are specified to meet operating standards across a range of defined variations that are derived from the expected operating variations and manufacturing variations. Common variables include process, voltage, temperature, threshold voltage (V_t), and offsets between matched sensitive circuits. The

R.C. Aitken (✉)
ARM Research, San Jose, CA, USA
e-mail: rob.aitken@arm.com

A. Singhee and R.A. Rutenbar (eds.), *Extreme Statistics in Nanoscale Memory Design*, Integrated Circuits and Systems, DOI 10.1007/978-1-4419-6606-3_8,
© Springer Science+Business Media, LLC 2010

range of these variables in which circuit operation is guaranteed is called the operating range. In order to be confident that a manufactured circuit will function properly across its anticipated operating range (where conditions will not be as ideal as they are in simulation), it is necessary to stress variations beyond the operating range. The difference between the design space defined by the stress variations and the operating variations is called "margin," as shown in Fig. 8.1.

Increased design margin generally translates into improved product yield, but the relationship is complex. Yield is composed of three components: random (defect-related) yield, systematic (design-dependent) yield, and parametric (process-related) yield. Electrical design margin translates into a lower bound on systematic and parametric yield, but does not affect defect-related yield. Most layout margins (adjustments for manufacturability) translate into improved random yield, although some improvements in systematic and parametric yield also occur.

It is useful to distinguish the concept of being outside the margin range of a memory ("*margin yield*") from its "true" yield. The act of setting design margins with a known occurrence probability places a lower bound on silicon yield – devices whose parameters are no worse than the margin case that will operate correctly. Devices on the other side are not checked, so the bound is not tight. In other words, it is possible for actual silicon with parameters outside margin space to function correctly, but it is not guaranteed (and, depending on the parameter in question, may be very unlikely). Note also that inaccurate estimation of the likelihood of the margin point can lead to margin yield overestimating true yield. In addition, there is a relationship between test methodology and yield – devices may fail a test while working in an application or vice versa. The issues are beyond the scope of this chapter and the interested reader is referred to works such as [26] for more discussion. For simplicity, when we refer to "yield," we are referring to margin yield: the probability that a manufactured device will be within the *estimated* margin range. By extension, a "failure" is an instance that is outside this margin range. This is shown graphically in Fig. 8.2.

When multiple copies of an object Z (e.g., a bit cell) are present in a memory, it is important for margin purposes to cover the worst-case behavior of any instance of

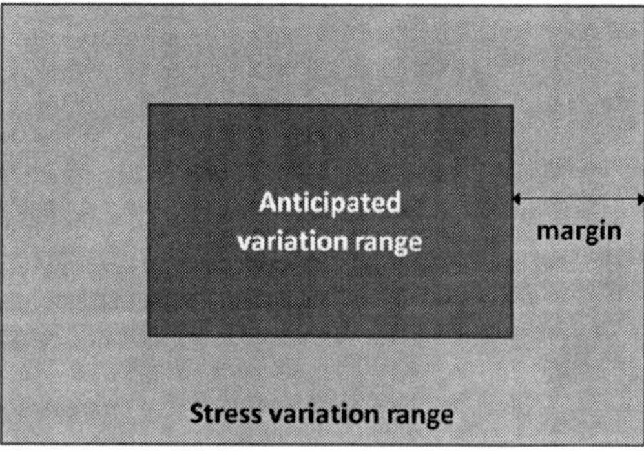

Fig. 8.1 Design margin

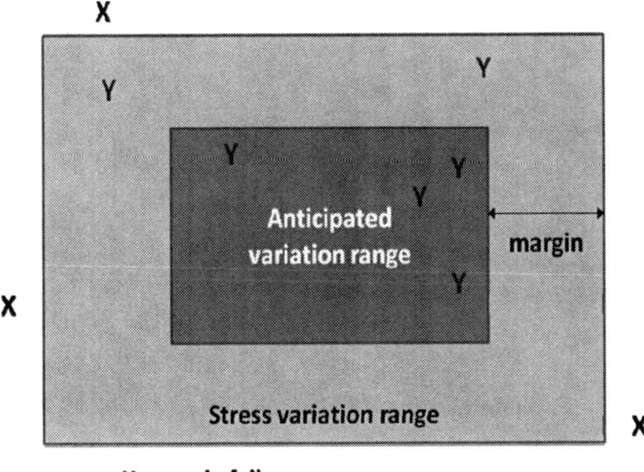

Fig. 8.2 Parametric variation leading to yielding instances and margin failures (instances outside margin range)

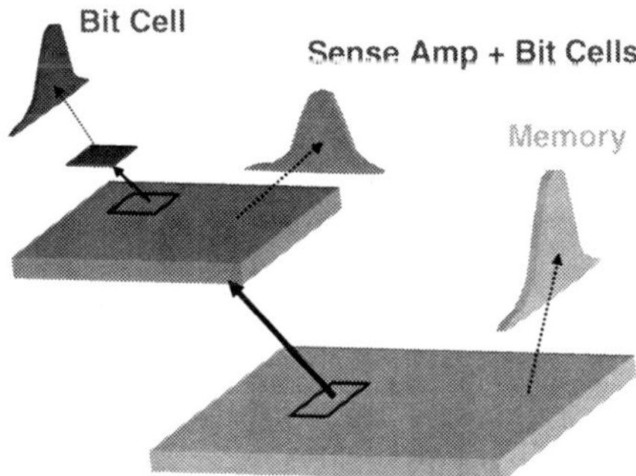

Fig. 8.3 System of variation within a memory

Z within the memory. For example, sense amplifier timing must accommodate the slowest bit cell (i.e., the cell with the lowest read current). The memory in turn must accommodate the worst-case combination of sense amp and bit cell variation, in the context of its own variation. The general issue is shown graphically in Fig. 8.3.

For design attributes that vary statistically, the worst-case value is itself covered by a distribution. Experimental evidence shows that read current has roughly a Gaussian (normal) distribution (see Fig. 8.4, and also [1]). The mean of this distribution depends on global process parameters (inter-die variation, whether the entire chip is "fast" or "slow" relative to others), and on chip-level variation (intra-die variation; e.g., across chip line-width variation). Margining is concerned with random (e.g. dopant) variation about this mean. Within each manufactured instance, individual bit cells can be treated as statistically independent of one another; that is, correlation between nearby cells, even adjacent cells, is minimal.

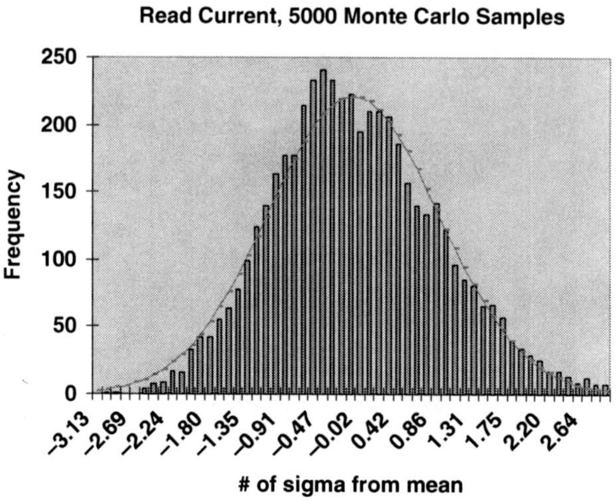

Fig. 8.4 Read current variation for a 65 nm bit cell

Every manufactured memory instance will have a bit cell with lowest read current. After thousands of die have been produced, a distribution of "weakest cells" can be gathered. This distribution will have a mean, the "expected worst-case cell," and a variance.

This chapter shows that setting effective margins for memories depends on carefully analyzing this distribution of worst-case values. It will be shown that the distribution is not Gaussian, even if the underlying data are Gaussian, and that effective margin setting requires separate treatment of bit cells, sense amplifiers, and self-timing paths.

One key factor in any discussion on the statistics of memory margining is that the events of greatest interest are those in the tails of the relevant distributions. By definition, these are rare events, yet the large numbers of bit cells make these rare events common enough to dominate the statistical analysis of memory functionality and yield.

The remainder of the chapter discusses the application of rare event statistics and EVT to memory design and margining. The chapter is organized as follows. Section 8.2 explains the concepts of extreme values and tails, and reviews some useful results from EVT that suggest appropriate limiting distributions for extreme values. Section 8.3 discusses some nonparametric techniques for analyzing extreme values, in particular order statistics, quantiles and the mean excess plot. Section 8.4 describes some methods of parametric fitting of tail distributions. An efficient rare event sampling technique called statistical blockade is described in Sect. 8.5, which then uses the methods of Sect. 8.4 to generate complete probability distributions of the relevant rare events. Section 8.6 describes some easy-to-use, yet statistically rigorous margining methods based on the distribution of worst-case values. The section shows how these methods can be applied at the array level by considering the statistics of the bit cell and the sense circuitry simultaneously.

8.2 Extremes: Tails and Maxima

Suppose we want to model the rare event statistics of the write time of an SRAM cell. Figure 8.5 shows an example of the distribution of the write time.

We see that it is skewed to the right with a *heavy* right tail. A typical approach is to run a Monte Carlo with a small sample size (e.g., 1,000) and fit a standard analytical distribution to the data, for example, a normal or a lognormal distribution. Such an approach can be accurate for fitting the "body" of the distribution, but will often be grossly inaccurate in the tail of the distribution: the skewness of the actual distribution or the heaviness of its tail will be difficult to match. As a result, any prediction of the statistics of rare events, lying far in the tail, will be inaccurate.

Let F denote the cumulative distribution function (CDF) of the write time y, and let us define a tail threshold t to mark the beginning of the tail (e.g., the 99th percentile). Let z be the excess over the threshold t. We can write the *conditional* CDF of the tail as

$$F_t(z) = P(Y - t \le z | Y > t) = \frac{F(z + t) - F(t)}{1 - F(t)}, \tag{8.1}$$

and the overall CDF as

$$F(z + t) = (1 - F(t))F_t(z) + F(t). \tag{8.2}$$

If we know $F(t)$ and can estimate the conditional CDF of the tail $F_t(z)$ accurately, we can accurately estimate rare event statistics. For example, the yield for some extreme threshold $y_f > t$ is given as

$$F(y_f) = (1 - F(t))F_t(y_f - t) + F(t), \tag{8.3}$$

and the corresponding failure probability $F_f(y_f) = 1 - F(y_f)$ is given as

$$F_f(y_f) = (1 - F(t))(1 - F_t(y_f - t)). \tag{8.4}$$

$F(t)$ can be accurately estimated using a few thousand simulations, since t is not too far out in the tail. Then, the problem here is to efficiently estimate the

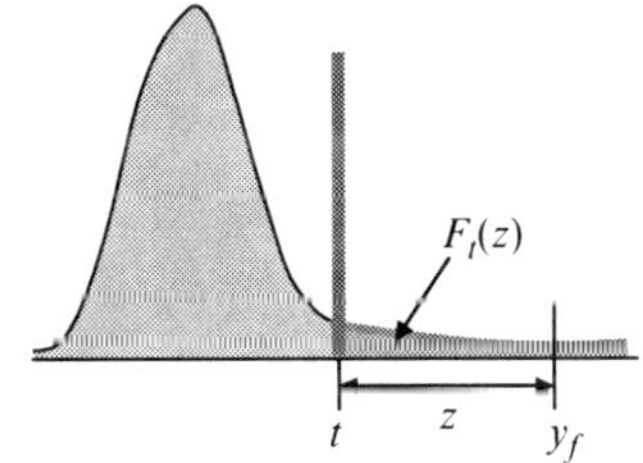

Fig. 8.5 A possible skewed distribution for some SRAM metric (e.g., write time)

conditional tail CDF F_t as a simple analytical form, which can then be used to compute statistical metrics, such as (8.3) and (8.4), for rare events. Of course, here we assume that any threshold y_f of interest will be far into the tail, such that $y_f \gg t$. This is easily satisfied for any real memory design scenario, since memory sizes are typically >1 Mb, requiring failure probabilities on individual bit cells $\ll$ 1 ppm in order to achieve overall memory yields in an acceptable range (typically > 99%).[1] We also assume that the extreme values of interest lie only in the *upper* tail of the distribution. This is without any loss of generality because any lower tail can be converted to the upper tail by replacing $y = -y$. This same approach of fitting a CDF to the exceedances over some threshold has been developed and widely applied by hydrologists under the name of the *peaks over threshold* (POT) method [2]. In their case though, the data is from historical record and not synthetically generated.

Now, suppose that Y_1, Y_2, ... is a sequence of independent, identically distributed random variables from the CDF F. For any sample Y_1, Y_2, ..., Y_N of size N, define the *sample maximum* as

$$M_N = \max(Y_1, Y_2, \ldots, Y_N), \quad N \geq 2. \tag{8.5}$$

The probability of $M_N \leq y$ is the probability of all of Y_1, Y_2, ..., Y_N being $\leq y$, as in:

$$P(M_N \leq y) = P(Y_1 \leq y, \ldots, Y_N \leq y) = \Pi_{i=1}^{N} P(Y_i \leq y) = F^N(y). \tag{8.6}$$

The maximum is a measure of the expected worst-case value of any performance metric. Consequently, the statistics of the maximum can help us analyze the statistical worst-case behavior of a circuit and estimate accurate statistical design margins.

We now look at some results from EVT that directly apply to these problems of estimating rare event statistics: the sample maxima and the conditional tail CDF.

8.2.1 Sample Maximum: Limiting Distributions

An important result from EVT addresses the question: *What are the possible limiting distributions of M_N as $N \to \infty$?* This result is stated in the following theorem by Fisher and Tippett [3].

---

[1] To put this in perspective, note that achieving 99% yield for even 1,000 bit cells requires marginal yield of 99.999% on each cell. For a parameter with an unskewed Gaussian distribution, this equates to over 4.2 standard deviations beyond the mean, far above the 2.3 assumed by a 99th percentile value. With megabits of memory, it is clear that assuming thresholds far into the tail is reasonable

Theorem 8.1 *(Fisher–Tippett [3]). If there exist normalizing constants a_N, b_N, and some nondegenerate CDF H, such that*

$$P\left(\frac{M_N - b_N}{a_N} \leq y\right) = F^N(a_N y + b_N) \to H(y) \quad \text{as } N \to \infty, y \in \Re, \tag{8.7}$$

then H belongs to the type of one of the following three CDFs:

$$\text{Frechet:} \quad \Phi_\alpha(y) = \begin{cases} 0, & y \leq 0 \\ e^{-y^{-\alpha}}, & y > 0 \end{cases}, \quad \alpha > 0,$$

$$\text{Weibull:} \quad \Psi_\alpha(y) = \begin{cases} e^{-(-y)^\alpha}, & y \leq 0 \\ 1, & y > 0 \end{cases}, \quad \alpha > 0, \tag{8.8}$$

$$\text{Gumbel:} \quad \Lambda(y) = e^{-e^{-y}}, \quad y \in \Re.$$

This amazing result formed the foundation of estimation of rare event statistics. Roughly, it says that for a very large class of CDFs, we can model the distribution of the normalized sample maximum M_N as one of three standard distributions: Fréchet, Weibull, and Gumbel. These three CDFs can be combined together into a *generalized extreme value* (GEV) distribution:

$$H_\xi(y) = \begin{cases} e^{-(1-\xi y)^{1/\xi}}, & \xi \neq 0, \\ e^{-e^{-y}}, & \xi = 0, \end{cases} \quad \text{where } 1 - \xi y > 0 \tag{8.9}$$

The three CDFs are obtained as follows.

- $\xi = -\alpha^{-1} < 0$ gives the Fréchet CDF Φ_α,
- $\xi = \alpha^{-1} > 0$ gives the Weibull CDF Ψ_α, and
- $\xi = 0$ gives the Gumbell CDF Λ.

The condition (8.7) is commonly stated as *F lies in the maximum domain of attraction of H*, or $F \in \text{MDA}(H)$. Hence, for nondegenerate H, Theorem 8.1 can be stated succinctly as

$$F \in \text{MDA}(H) \Rightarrow H \quad \text{is of type } H_\xi.$$

The conditions for which $F \in \text{MDA}(H)$ for some nondegenerate H are quite general for most practical purposes and known well. Gnedenko [4] provided the first rigorous proof for the Fisher–Tippett theorem, showing conditions on F required for the convergence to each of the three limiting CDFs. For details regarding these conditions, see [2, 5, 6]. Here, we only list some common distributions belonging to $\text{MDA}(H_\xi)$, in Table 8.1, and immediately proceed to a similar result for the conditional CDF of the tail.

Table 8.1 Some common distributions lying in MDA(H_ξ)

H_ξ	Distributions in MDA(H_ξ)
$\Phi_{-1/\xi}$	Cauchy
	Pareto
	Loggamma
$\Psi_{1/\xi}$	Uniform
	Beta
Λ	Normal
	Lognormal
	Gamma
	Exponential

8.2.2 Distribution Tail: Limiting Distributions

We recall the definition of F_t as the conditional tail CDF for a tail threshold t, as in (8.1). Then, the following is true.

Theorem 8.2 *(Balkema and de Haan [7] and Pickands [8]).*
For every $\xi \in \Re$, $F \in MDA(H_\xi)$ if and only if

$$\lim_{t \to \infty} \sup_{z \geq 0} \left| F_t(z) - G_{\xi,\beta(t)}(z) \right| = 0 \tag{8.10}$$

for some positive function $\beta(t)$, where $G_{\xi,\beta}(z)$ is the generalized Pareto distribution (GPD)

$$G_{\xi,\beta}(z) = \begin{cases} 1 - (1 - \xi\frac{z}{\beta})^{1/\xi}, & \xi \neq 0, z \in D(\xi,\beta), \\ 1 - e^{-z/\beta}, & \xi = 0, z \geq 0, \end{cases}.$$

where

$$D(\xi,\beta) = \begin{cases} [0,\infty), & \xi \leq 0, \\ [0,\beta/\xi], & \xi > 0. \end{cases} \tag{8.11}$$

In other words, for any distribution F in the maximum domain of attraction of the GEV distribution, the conditional tail distribution F_t converges to a GPD as we move further out in the tail. This is an extremely useful result: it implies that, if we can generate enough points in the tail of a distribution ($y \geq t$), in most practical cases, we can fit the simple, analytical GPD to the data and make predictions further out in the tail. This approach would be independent of the circuit or the performance metric being considered.

Of course, two important questions remain:

- How do we efficiently generate a large number of points in the tail ($y \geq t$)?
- How do we fit the GPD to the generated tail points?

We answer these in the following sections.

8.3 Analysis of Tails and Extreme Values

In this section, we look at some useful statistics and techniques for analyzing and understanding the behavior of extreme values.

8.3.1 Order Statistics and Quantiles

Monte Carlo based methods or statistical measurements yield samples of some quantity, for instance, the write time of an SRAM cell. Each sampled value is then a random variable. It is often instructive to study the quantiles and order statistics of the sample. For instance, the median is the 50% quantile (or 50th percentile) of the distribution, and is often used as a robust measure of the most common case. The sample maximum M_N from (8.5) is a measure of the 100% quantile.

8.3.1.1 Order Statistics

Let us start with order statistics first. Once again, let Y_1, Y_2, ..., Y_N be the i.i.d. random variables drawn from the CDF F. First, we define the *order statistics* for such a sample. If we sort the sample in increasing order, as

$$Y_{1,N} \leq Y_{2,N} \leq \cdots \leq Y_{N,N},$$

to obtain the *ordered sample*, then $Y_{k,N}$ is the *kth order statistic*. Note that $Y_{1,N} = \min(Y_1, \ldots, Y_N)$, and $Y_{1,N} = \max(Y_1, \ldots, Y_N)$ are the sample minimum and maximum, respectively. $Y_{k,N}$ is also referred to as the $(N - k + 1)$th *upper* order statistic. We can easily infer an *empirical CDF* as an approximation for F from the order statistics. For any real y, we write this empirical CDF, $F_N(y)$, as

$$F_N(y) = \frac{1}{N} \left| \{ Y_{k,N} : Y_{k,N} \leq y, \ 1 \leq k \leq N \} \right|;$$

that is, the fraction of sample values that are less than y. Figure 8.6 shows such an empirical CDF computed using a sample from a lognormal distribution.

The distribution of the kth order statistic drawn from the CDF F can be easily derived as follows. We denote the CDF of the kth order statistic by $F_{k,N}(y)$. Then, we can write

$$\begin{aligned}
F_{k,N}(y) &= P(Y_{k,N} \leq y) \\
&= P(Y_{k,N} \leq y{<}Y_{k+1,N}) + P(Y_{k+1,N} \leq y{<}Y_{k+2,N}) + \cdots + P(Y_{N,N} \leq y) \\
&= {}^N C_{N-k}(1 - F(y))^{N-k} F(y)^k + \cdots + {}^N C_0(1 - F(y))^0 F(y)^N \\
&= \sum_{r=0}^{N-k} {}^N C_r (1 - F(y))^r F(y)^{N-r}.
\end{aligned}$$

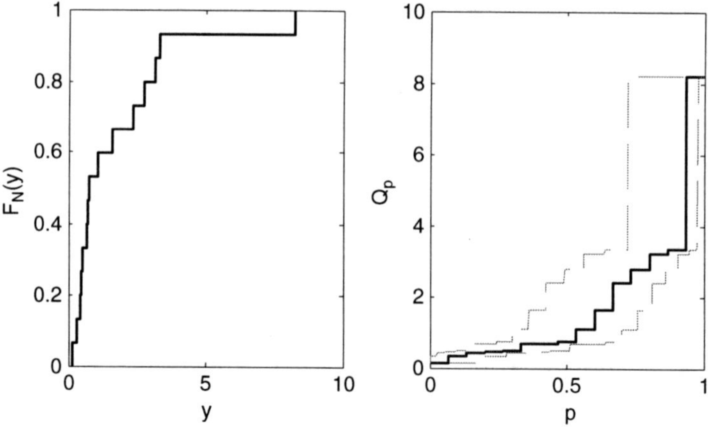

Fig. 8.6 Empirical CDF and sample quantiles estimated for a sample of 15 values drawn from a lognormal distribution. The quantile plot also shows the 90% confidence intervals, estimated using (8.13)

8.3.1.2 Quantiles

The Pth percentile ($0 < P < 100$) is that value of y that has a cumulative probability of $P/100$. Hence, we divide the entire domain of y into 100 equiprobable bins, and the percentiles are the bin boundaries. Quantiles are generalizations of this concept. Instead of 100 bins, if we make q equiprobable bins, then the resulting boundary values are the q-quantiles. For $0 < k < q$, and integer k and q, the kth q-quantile or just kth q-tile is the smallest value of y that has a cumulative probability no less than k/q.

We can further generalize this idea to the p probability quantile, where $0 < p < 1$. The *quantile* for probability p is the smallest value y that has a cumulative probability no less than p. If we denote this quantity as Q_p, then we can define it in terms of the *quantile function $F^{-1}(p)$*:

$$Q_p = F^{-1}(p) = \inf\{y \in \Re : F(y) \geq p\}, \quad 0 < p < 1.$$

Quantiles can be estimated from sampled data, using the order statistics. A common estimate of the *sample quantile* for probability p is given as

$$Q_{p,N} = F_N^{-1}(p) = Y_{k,N} : \frac{k-1}{N} < p \leq \frac{k}{N}, \quad k = 1, \dots, N. \tag{8.12}$$

This expression lets us compute, for instance, the 99th percentile of the write time of an SRAM cell from some sampling of the write time: $p = 0.99$. Figure 8.6 shows the empirical quantile function for a sample of lognormal random variables. An important question now is: *What is the distribution of the sample quantile?* If we know this distribution, we can estimate confidence intervals for any quantile estimate. We address this question next.

Suppose we have estimated some quantile $Q_{p,N}$ using an ordered sample. An empirical distribution for Q_p can be estimated using the ordered sample as follows.

$$F_{Q_p,N}(y) = \begin{cases} P(Q_p \leq Y_{k,N}) : k = \sup\{i : Y_{i,N} \leq y\}, & y \geq Y_{1,N}, \\ 0, & y < Y_{1,N}. \end{cases}$$

Now,

$$\begin{aligned} P(Q_p \leq Y_{k,N}) &= P(Q_P \leq Y_{1,N}) + \cdots + P(Y_{k-1,N} < Q_P \leq Y_{k,N}) \\ &= {}^N C_0 p^0 (1-p)^{N-0} + \cdots + {}^N C_{k-1} p^{k-1} (1-p)^{N-k+1} \\ &= \sum_{r=0}^{k-1} {}^N C_r p^r (1-p)^{N-r} \\ &= B(k-1; N, p), \end{aligned}$$

where $B(\bullet; N, p)$ is the Binomial distribution with parameters N and p. This gives us a convenient estimate for the confidence interval of any quantile estimate. Suppose we wish to compute *P% confidence limits* for the quantile estimate $Q_{p,N}$ computed using (8.12). We can use the following estimate:

$$[Y_{l,N}, Y_{h,N}], \quad \text{where}$$
$$l = B^{-1}(0.5 - P/200; N, p) + 1, \quad h = B^{-1}(0.5 + P/200; N, p) + 1. \qquad (8.13)$$

Here, $Y_{l,N}$ and $Y_{h,N}$ are order statistics. The confidence interval estimates for our lognormal example are shown in Fig. 8.6.

For large sample cases, the binomial distribution $B(y,N,p)$ is well approximated by a normal distribution with mean Np and variance $Np(1-p)$. Hence, if N is large, we can estimate the indices l and h in (8.13) as

$$l = Np - \Delta, \quad h = Np + \Delta, \quad \text{where } \Delta = \Phi^{-1}(0.5 + P/200)\sqrt{Np(1-p)},$$

where Φ^{-1} is the inverse standard normal CDF.

8.3.2 Mean Excess Plot

A common tool for graphical exploration of statistical data, particularly of distribution tails, is the *sample mean excess plot*, an idea introduced by Davison and Smith [9]. It is particularly useful for identifying the GPD from sampled tails. The *mean excess function* for a given threshold t is defined as

$$e(t) = E(y - t | y > t);$$

that is, the mean of exceedances over t. Plotting $e(t)$ against t gives us the mean excess plot. The *sample* mean excess function is the sample version of $e(t)$. For a given sample $\{y_i: i = 1,\ldots,n\}$, it is defined as

$$e_n(t) = \frac{\sum_{i=1}^{n}(y_i - t)_+}{|\{y_i : y_i > t\}|}, \quad \text{where } (\bullet)_+ = \max(\bullet, 0);$$

that is, the sample mean of only the exceedances over t. A plot of $e_n(t)$ against t gives us the sample mean excess plot. The mean excess function of a GPD $G_{\xi,\beta}$ can be shown (*see* [2]) to be a straight line given by

$$e(t) = \frac{\beta - \xi t}{1 + \xi} \quad \text{for} \quad t \in D(\xi, \beta), \tag{8.14}$$

where $D(\xi, \beta)$ is as defined in Theorem 8.2. Hence, if the sample mean excess function of any data sample starts to follow roughly a straight line from some threshold, then it is an indication that the exceedances over that threshold follow a GPD.

Suppose we have some sample data $Y_1, \ldots, Y_n$, and we wish to identify if and at what value (tail threshold) the GPD convergence sets in. We can exploit the mean excess plot to test the GPD convergence of any tail threshold we choose. Let us look at an example to illustrate these ideas.

Let us generate a large sample from the standard normal distribution, and fit GPDs at two tail thresholds: $t = 1$ and $t = 2.5$. We use the maximum likelihood estimation (MLE) method, as described in the following sections, to obtain these fits. The resulting mean excess relations from (8.14) are then compared against the sample mean excess plot of the normal sample. This comparison is shown in Fig. 8.7. It is obvious that the GPD behavior has not set in at $t = 1$, since the sample mean excess plot does not follow the straight line mean excess relation of the GPD. However, going further out into the tail, at $t = 2.5$, it seems that the GPD behavior has indeed emerged.

8.4 Estimating the Tail: Fitting the GPD to Data

For now, let us suppose that we can generate a reasonably large number of points in the tail of our performance distribution. For this, we might theoretically use standard Monte Carlo simulation with an extremely large sample size, or, more practically, we discuss the statistical blockade sampling method in Sect. 8.5. Let this data be $\mathbf{Z} = (Z_1,\ldots, Z_n)$, where each Z_i is the *exceedance* over the tail 'threshold t $(Z_i > 0$, for all i). All Z_i are i.i.d. random variables with common CDF F_t. Then we have the problem of estimating the optimal GPD parameters ξ, β from this tail data, so as to best fit the conditional tail CDF F_t. There are several options; we review three of the most popular ones here. In particular, we focus on two methods that require no manual effort and can be completely optimized.

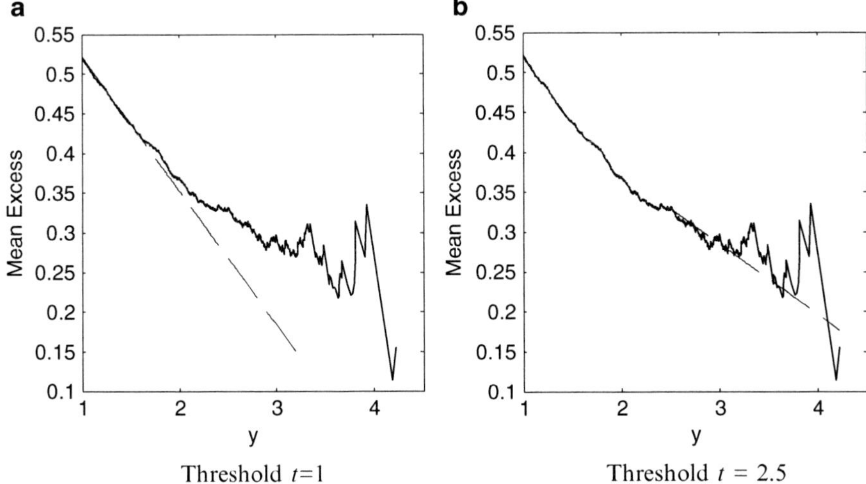

Fig. 8.7 Sample mean excess plot for a sample drawn from the standard normal. The GPD behavior of linear mean excess plot appears to have set in beyond a tail threshold of 2.5

8.4.1 Maximum Likelihood Estimation

MLE is a standard statistical estimation technique that tries to estimate those model parameter values (here ξ, β of the GPD) that maximize the "chances" of obtaining the data that we have observed. The probability density function (PDF) of a GPD $G_{\xi,\beta}$ is given as

$$g_{\xi,\beta}(z) = \begin{cases} \frac{1}{\beta}\left(1 - \xi\frac{z}{\beta}\right)^{1/\xi-1}, & \xi \neq 0,\; z \in D(\xi,\beta) \\ \frac{1}{\beta}e^{-z/\beta}, & \xi = 0,\; z \geq 0 \end{cases}, \tag{8.15}$$

where $D(\xi, \beta)$ is defined in Theorem 8.2. Recall that all Z_i are i.i.d. random variables with common CDF F_t. We assume that F_t is of the form of a GPD. The *likelihood* ("chances") of having seen this data from an underlying GPD is the multivariate probability density associated with it, and is given as

$$\ell(\xi,\beta|Z) = g_{\xi,\beta}(Z_1,\ldots,Z_n) = \prod_{i=1}^{n} g_{\xi,\beta}(Z_i).$$

Since $\ell(\xi, \beta \mid Z)$ can be too small for accurate computation with finite accuracy, it is typical to use the *log-likelihood function*

$$\ln(\ell(\xi,\beta|Z)) = \sum_{i=1}^{n} \ln\big(g_{\xi,\beta}(Z_i)\big),$$

which increases monotonically with ℓ. MLE then computes (ξ, β) to maximize this log-likelihood, as

$$(\hat{\xi}, \hat{\beta})_{mle} = \arg\max_{\xi,\beta} \sum_{i=1}^{n} \ln\big(g_{\xi,\beta}(Z_i)\big). \tag{8.16}$$

Substitution of (8.15) in (8.16) and subsequent algebra allows for a simplification to a one dimensional search that can be exploited by a careful implementation of a Newton–Raphson algorithm. Details regarding such an implementation are shown in [10].

Smith [11] studies convergence when F_t is not exactly of GPD form, and provides limit results for the distributions of $(\hat{\xi}, \hat{\beta})_{\text{mle}}$ for each of the three cases, $F \in \text{MDA}(\Phi_{-1/\xi})$, $F \in \text{MDA}(\Lambda)$, and $F \in \text{MDA}(\Psi_{1/\xi})$. For $\xi < 1/2$, the MLE estimates are asymptotically normal and efficient (bias $= 0$) under certain regularity assumptions on F. If (ξ, β) are the exact values to be estimated, then as the sample size $n \to \infty$, the variance of the MLE estimates is given as

$$\text{var}\begin{bmatrix} \hat{\xi} \\ \hat{\beta} \end{bmatrix} \to \frac{1-\xi}{n} \begin{bmatrix} 1-\xi & \beta \\ \beta & 2\beta^2 \end{bmatrix}, \quad \xi < \frac{1}{2}.$$

When $\xi \geq 1/2$, MLE convergence can be difficult and special techniques are needed, as discussed in [12]. However, $\xi \geq 1/2$ is usually rare, since it corresponds to a finite tail with $g_{\xi,\beta}(z) > 0$ at the endpoint (Fig. 8.8).

8.4.2 Probability-Weighted Moment Matching

Probability-weighted moments (PWMs) [13] of a continuous random variable Y with CDF F are generalizations of the standard moments, and are defined as

$$M_{p,r,s} = E[Y^p F^r(Y)(1 - F(Y))^s].$$

The standard pth moment is given by $M_{p,0,0}$. For the GPD, we have a convenient relationship between $M_{1,0,s}$ and (ξ, β), given by

$$m_s = M_{1,0,s} = \frac{\beta}{(1+s)(1+s+\xi)}, \quad \xi > 0.$$

Then, we can write

$$\beta = \frac{2m_0 m_1}{m_0 - 2m_1}, \quad \xi = \frac{m_0}{m_0 - 2m_1} - 2.$$

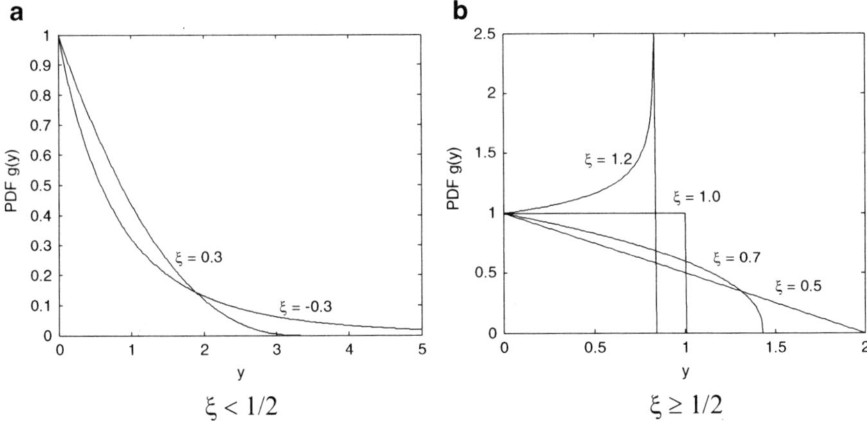

Fig. 8.8 The probability density function for a GPD with $\beta = 1$. We get long unbounded tails for $\xi \leq 0$

We estimate these PWMs from the data sample, as

$$\hat{m}_s = \frac{1}{n} \sum_{i=1}^{n} (1 - q_i)^s Y_{i,n},$$

where $Y_{1,n} \leq Y_{2,n} \leq \ldots \leq Y_{n,n}$ is the *ordered sample*, and

$$q_i = \frac{i + \gamma}{n + \delta}.$$

Here, γ and δ are fitting parameters. We can use $\gamma = -0.35$ and $\delta = 0$, as suggested for GPD fitting in [14]. The estimates $\hat{\xi}, \hat{\beta}$ converge to the exact values as $n \to \infty$, and are asymptotically normally distributed with covariance given by

$$\text{var} \begin{bmatrix} \hat{\xi} \\ \hat{\beta} \end{bmatrix} \to \frac{n^{-1}}{(1 + 2\xi)(3 + 2\xi)}$$

$$\times \begin{bmatrix} (1 + \xi)(2 + \xi)^2(1 + \xi + 2\xi^2) & \beta(2 + \xi)(2 + 6\xi + 7\xi^2 + 2\xi^3) \\ \beta(2 + \xi)(2 + 6\xi + 7\xi^2 + 2\xi^3) & \beta^2(7 + 18\xi + 11\xi^2 + 2\xi^3) \end{bmatrix}$$

$$\tag{8.17}$$

Based on an extensive simulation study, [14] suggests that the PWM method often has lower bias than moment matching and MLE for sample sizes up to 500. Also, the MLE search (8.16) is shown to suffer from some convergence problems when ξ is estimated close to 1/2. Finally, the study also suggests that PWM matching gives more reliable estimates of the variability of the estimated parameters, as per (8.17). Based on these reasons, we choose PWM matching here.

Once we have estimated a GPD model of the conditional CDF above a threshold t, we can estimate the failure probability for any value y_f by substituting the GPD in (8.4) as

$$P(Y > y_f) \approx (1 - F(t))(1 - G_{\xi,\beta}(y_f - t)). \tag{8.18}$$

The next section addresses the important remaining question: How do we efficiently generate a large number of points in the tail ($y \geq t$)?

8.5 Statistical Blockade: Sampling Rare Events

Let any circuit performance metric, or simply, output y be computed as

$$y = f_{\text{sim}}(\mathbf{x}). \tag{8.19}$$

Here, $\mathbf{x}$ is a point in the statistical parameter (e.g., V_t, t_{ox}) space, or simply, the input space, and f_{sim} includes expensive SPICE simulation. We assume that y has some probability distribution F, with an extended tail. Suppose we define a large tail threshold t for y, then from the developments in Sect. 8.2 we know that we can approximate the conditional tail CDF F_t by a GPD $G_{\xi,\beta}$. Section 8.4 shows how we can estimate the GPD parameters (ξ, β) from data drawn from the tail distribution. We now discuss an efficient tail sampling strategy that will generate the tail points for fitting this GPD.

Corresponding to the tail of output distribution, we expect a "tail region" in the input space: any statistical parameter values drawn from this tail region will give an output value $y > t$. Figure 8.9 shows an example of such a tail region for two inputs. The rest of the input space is called the "body" region, corresponding to the body of the output distribution F. In Fig. 8.9 these two regions are separated by a dashed line.

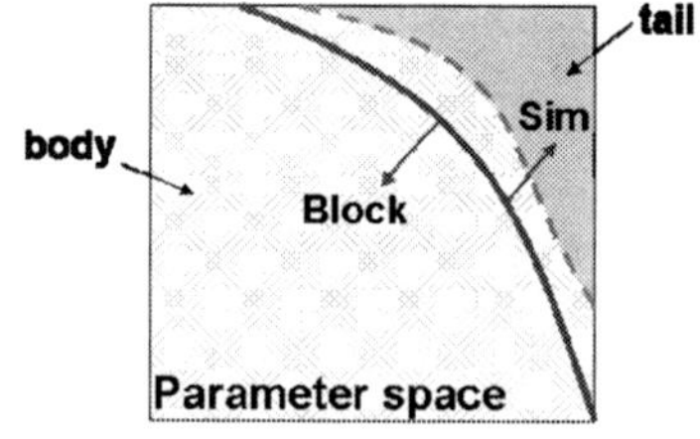

Fig. 8.9 The tail and body regions in the statistical parameter space. The *dashed line* is the exact tail region boundary for tail threshold t. The *solid line* is the relaxed boundary modeled by the classifier for a classification threshold $t_c < t$

The key idea behind the efficient sampling technique is to identify the tail region and simulate only those Monte Carlo points that are likely to lie in this tail region. Here, we exploit the common fact that *generating* the random values for a Monte Carlo sample point is very cheap compared to actually *simulating* the point as in (8.19). Hence, if we generate points as in standard Monte Carlo, but *block* — not simulate — those points that are unlikely to fall in the tail region, we can drastically cut down the total time spent. We use a classifier to distinguish between the tail and body regions and to block out the body points. A classifier [15] is an indicator function that takes as input any point in the input space (the statistical parameter space) and predicts the membership of this point in one of the multiple classes (the "body" or "tail" classes). In the context of Fig. 8.9, it essentially builds a model of the boundary between the tail and body regions. Using this model of the boundary, it can label points from the Monte Carlo sample set as either "tail" or "body". Only the "tail" points are then simulated. We also refer to this classifier as the *blockade filter* and its blocking activity as *blockade filtering*.

To build this model of the tail region boundary, the classifier can be trained with a small (e.g., 1,000 points) training set of simulated Monte Carlo sample points. However, it is difficult, if not impossible to build an *exact* model of the boundary in general. Misclassifications, at least on points unseen during training, is unavoidable. Hence, we relax the accuracy requirement to allow for classification error. This is done by building the classification boundary at a *classification threshold* t_c that is less than the tail threshold t. Since we have assumed that only the upper (right) tail is relevant, the tail region corresponding to t will be a subset of the tail region corresponding to t_c, if $t_c < t$. This will help to ensure that, even if the classifier is imperfect, it is unlikely that it will misclassify points in the true tail region (for t). The relaxed boundary corresponding to such a t_c is shown as the solid line in Fig. 8.9. The statistical blockade algorithm is then as in Algorithm 8.1.

The thresholds $t = p_t$th percentile and $t_c = p_c$th percentile are estimated from the small initial Monte Carlo run, which also gives the n_0 training points for the classifier. Typical values for these constants are shown in Algorithm 8.1. The function *MonteCarlo(n)* generates n points in the statistical parameter space, which are stored in the $n \times s$ matrix $\mathbf{X}$, where s is the input dimensionality. Each row of $\mathbf{X}$ is a point in s dimensions. $\mathbf{y}$ is a vector of output values computed from simulations. The function *BuildClassifier*$(\mathbf{X}, \mathbf{y}, t_c)$ trains and returns a classifier using the training set $(\mathbf{X}, \mathbf{y})$ and classification threshold t_c. The function *Filter*$(C, \mathbf{X})$ blocks the points in $\mathbf{X}$ classified as "body" by the classifier C, and returns only the points classified as "tail." *FitGPD*$(\mathbf{y}_{\text{tail}} - t)$ computes the parameters (ξ, β) for the best GPD approximation $G_{\xi, \beta}$ to the conditional CDF of the exceedances of the tail points in $\mathbf{y}_{\text{tail}}$ over t. We can then use this GPD model to compute statistical metrics for rare events, for example, the failure probability for some threshold y_f, as in (8.18). This sampling procedure is also illustrated in Fig. 8.10.

$$\frac{\gamma_t}{\gamma_b} = \frac{\text{Number of 'body' points}}{\text{Number of 'tail' points}}.$$

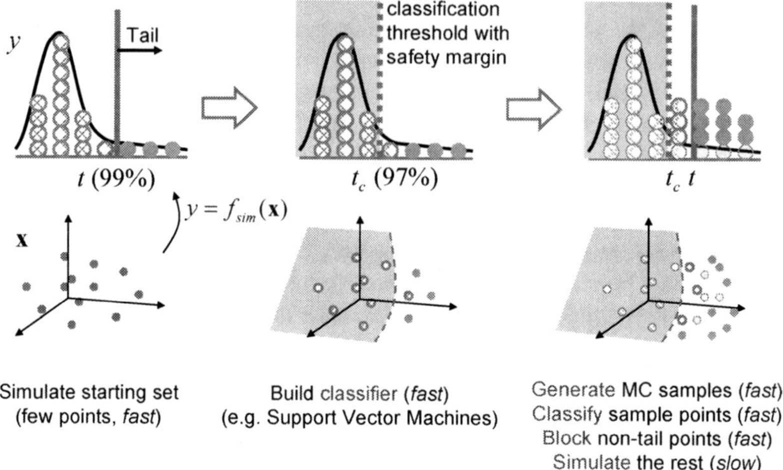

Fig. 8.10 The efficient tail (rare event) sampling method of statistical blockade

Require: training sample size n_0 (e.g., 1,000) total sample size n percentages p_t (e.g., 99%), p_c (e.g., 97%)	
1.	$\mathbf{X} = \text{MonteCarlo}(n_0)$
2.	$\mathbf{y} = f_{sim}(\mathbf{X})$
3.	$t = \text{Percentile}(\mathbf{y}, p_t)$
4.	$t_c = \text{Percentile}(\mathbf{y}, p_c)$
5.	$C = \text{BuildClassifier}(\mathbf{X}, \mathbf{y}, t_c)$ // C is a classifier
6.	$\mathbf{y} = f_{sim}(\text{Filter}(C, \text{MonteCarlo}(n)))$
7.	$\mathbf{y}_{tail} = \{y_i \in \mathbf{y} : y_i > t\}$
8.	$(\xi, \beta) = \text{FitGPD}(\mathbf{y}_{tail} - t)$

Algorithm 8.1. *The statistical blockade algorithm for efficiently sampling rare events and estimating their probability distribution.*

8.5.1 Unbiasing the Classifier

An important technical point to note about the classifier construction is as follows. The training set will typically have many more body points than tail points. Hence, even if all or most of the tail points are misclassified, the training error will be low as long as most of the body points are correctly classified. This will result in a classifier that is biased to allow more misclassifications of points in the tail region. However, we need to minimize misclassification of tail points to avoid distorting the statistics of the simulated tail points. Hence, we need to reverse bias the classification error.

Using the technique proposed in [16], we penalize misclassifications of tail points more than misclassifications of body points. Let γ_t and γ_b be possibly different penalty factors for the "tail" and "body" classes: misclassifications of any training points in the tail (body) region are penalized by a factor of γ_t (γ_b). We can obtain an unbiased classifier if we choose

$$\frac{\gamma_t}{\gamma_b} = \frac{\text{Number of 'body' points}}{\text{Number of 'tail' points}}.$$

8.5.2 Note on Sampling and Commercially Available Device Models

It is common for commercially available device models to include Monte Carlo variants. Care must be taken when using these to ensure that tails are being modeled correctly and that the sample space includes correct tail distributions. For example, many such models use the "AGAUSS" function, which generates a set of sample points within a set of bounds, typically set to $\pm 3\sigma$. Such an approach can produce some rare events (e.g., two independent parameters each at $+3\sigma$ is an event of about 2 ppm probability) but miss others that are much more likely (e.g., one parameter being above $+3.5\sigma$ while the other is less than 1σ occurs at a rate of 200 ppm). For this reason, when working with EVT, it is usually desirable to be very explicit about how sample points are generated and how they are incorporated into circuit simulations. The mechanics of doing this as part of a simulation flow vary considerably among commercially available device models, even those from a single foundry. A good approach for unencrypted models is to substitute externally generated sample points for those used by the model, taking care to consider underlying physical realities (avoiding negative V_t for NFETs for example). For encrypted models, an approximation can be made through the use of offsets. For example, Fig. 8.11 shows a method of using a parameterized voltage source providing a voltage offset to mimic V_t variation in an NFET.

8.5.3 Example: 6 T SRAM Cell

Let us now apply the statistical blockade method to a 6 T SRAM cell to illustrate its use. The circuit is shown in Fig. 8.12. Here, we use the Cadence 90 nm Generic PDK library for the BSIM3V3 [17] device models, and model RDF on every transistor as independent, normally distributed V_t variation per transistor. The V_t standard deviation is taken to be $\sigma(V_t) = 5(WL)^{-0.5}$ V, where W and L are the

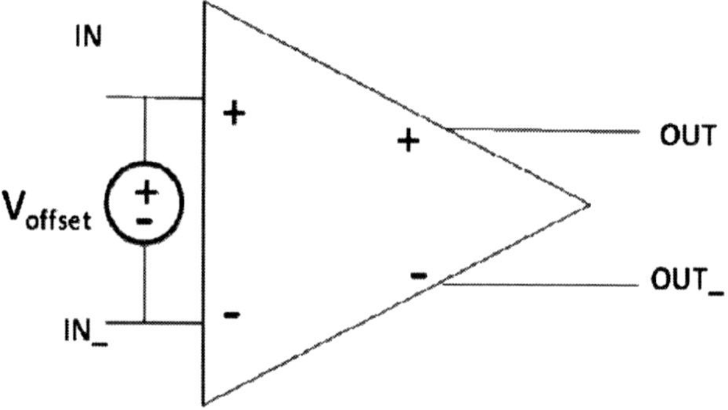

Fig. 8.11 Adding an offset to a sense amplifier to compensate for V_t shift on input pass transistors

Fig. 8.12 A 6-transistor
SRAM cell with write driver
and column mux

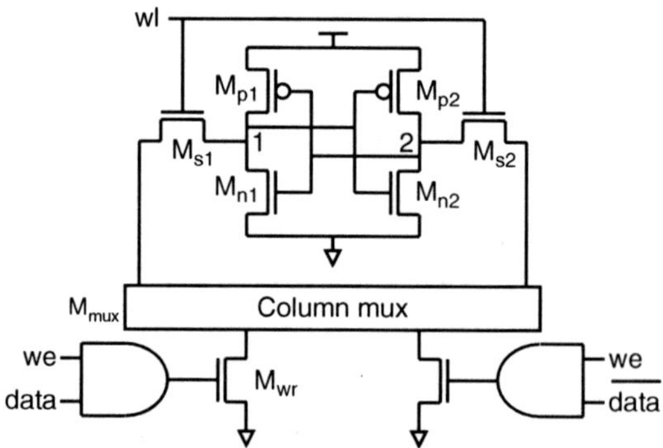

transistor width and length, respectively, in nm. We also include a global gate oxide
thickness variation, also normally distributed and with a standard deviation of 2%
of the nominal thickness. This gives us a total of 9 statistical parameters. The metric
we are measuring here is the write time τ_w: the time between the wordline going
high, to the nondriven cell node (node 2) transitioning. Here, "going high" and
"transitioning" imply crossing 50% of the full voltage change. The rare event
statistical metric that we compute is the failure probability $F_f(y_f)$ for any failure
threshold y_f. We represent this failure probability as the equivalent quantile y_σ on
the standard normal distribution:

$$y_\sigma = \Phi^{-1}(1 - F_f(y_f)) = \Phi^{-1}(F(y_f)),$$

where Φ is the standard normal CDF. For example, a failure probability of
$F_f = 0.00135$ implies a cumulative probability of $F = 1 - F_f = 0.99865$. The
equivalent point on a standard normal, having the same cumulative probability, is
$y_\sigma = 3$. In other words, any y_f with a failure probability of 0.00135 is a "3σ" point.

Let us now compute y_σ for different failure thresholds using statistical
blockade. The blockade filter is a Support Vector Machine classifier [18], built at

a classification threshold of $t_c = $ 97th percentile, using a training sample set of $n_0 = 1,000$ points. The tail threshold t is defined as the 99th percentile. We use (8.18), but the points used to estimate (ξ, β) are obtained from blockade filtering a sample of 100,000 points. This gives us 4,379 tail candidates, which on simulation yield 978 true tail points ($\tau_w > t$), to which the GPD is fit. We compare these results with a 1 million point Monte Carlo run, from which we estimate y_σ empirically.

Table 8.2 shows a comparison of the y_σ values estimated by the two different methods, and Fig. 8.13 compares the conditional tail CDFs computed from the empirical method and from statistical blockade, both showing a good match.

Some observations highlighting the efficiency of statistical blockade can be made immediately.

Table 8.2 Prediction of failure probability as y_σ by methods I, II and III, for a 6 T SRAM cell. The number of simulations for statistical blockade includes the 1,000 training points. The write time values are in "fanout of 4 delay" units

τ_w (y_f) (FO4)	(I) Standard Monte Carlo	(III) Statistical blockade
2.4	3.404	3.379
2.5	3.886	3.868
2.6	4.526	4.352
2.7	∞	4.845
2.8	∞	5.356
2.9	∞	5.899
3.0	∞	6.493
Num. Sims.	1,000,000	5,379

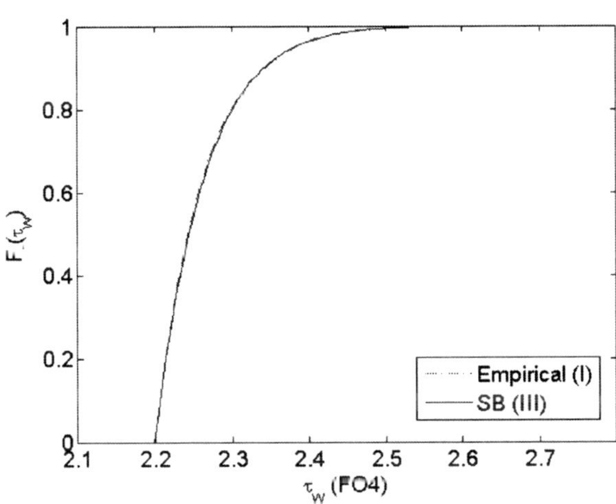

Fig. 8.13 Comparison of GPD tail model from statistical blockade (5,379 simulations) and the empirical tail CDF (one million simulations) for the write time of the 6 T SRAM cell

- The empirical method fails beyond 2.6 FO4, corresponding to about 1 ppm circuit failure probability because there are no points generated by the Monte Carlo run so far out in the tail.
- Fitting a GPD model to the tail points allows us to make predictions far out in the tail, even though we have no points that far out.
- Using blockade filtering, coupled with the GPD tail model, we can drastically reduce the number of simulations (from 1 million to 5,379) with very small change to the tail model.

However, the tail model cannot be relied on too far out from the available data because of the decrease in statistical confidence as we move further out in the tail. We further discuss and attack this problem in Sect. 8.5.5.

8.5.4 Conditionals and Disjoint Tail Regions

SRAM performance metrics are often computed for two states of the SRAM cell while storing a 1, and while storing a 0. The final metric value is then a maximum or a minimum of the values for these two states. The presence of such *conditionals* (max, min) can result in *disjoint* tail regions in the statistical parameter space, making it difficult to use a single classifier to define the boundary of the tail region. Let us look at an example to illustrate this problem.

Consider the 6 T SRAM cell. With technology scaling reaching nanometer feature sizes, subthreshold and gate leakage become very significant. In particular, for the large memory blocks seen today the standby power consumption due to leakage can be intolerably high. Supply voltage (V_{dd}) scaling [19] is a powerful technique to reduce this leakage, whereby the supply voltage is reduced when the memory bank is not being accessed. However, lowering V_{dd} also makes the cell unstable, ultimately resulting in data loss at some threshold value of V_{dd}, known as the *data retention voltage* or DRV. Hence, the DRV of an SRAM cell is the lowest supply voltage that still preserves the data stored in the cell. DRV can be computed as follows.

$$DRV = \max(DRV_0, \ DRV_1),\qquad(8.20)$$

where DRV_0 is the DRV when the cell is storing a 0, and DRV_1 is the DRV when it is storing a 1. If the cell is balanced (symmetric), with identical left and right halves, then $DRV_0 = DRV_1$. However, if there is any mismatch due to process variations, they become unequal. This creates the situation, where the standard statistical blockade classification technique would fail because of the presence of disjoint tail regions.

Suppose we run a 1,000-point Monte Carlo, varying all the mismatch parameters in the SRAM cell according to their statistical distributions. This would give us distributions of values for DRV_0, DRV_1, and DRV. In certain parts of the mismatch

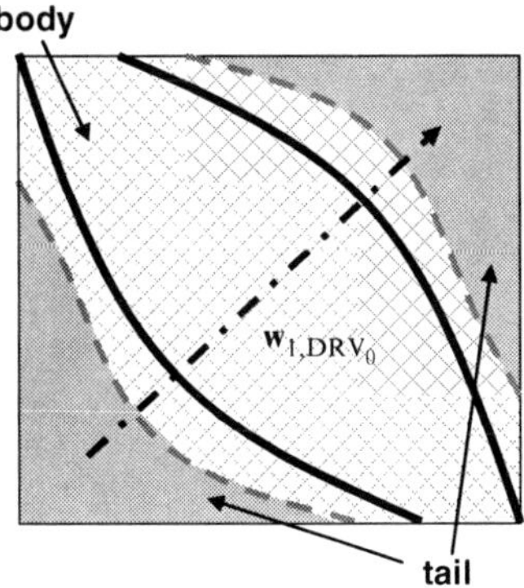

Fig. 8.14 A circuit metric (e.g., DRV) with two disjoint tail regions. The tail regions are separated from the body region by *dashed lines*. $w_{1,DRV0}$ is the direction of maximum variation of the circuit metric

parameter space $DRV_0 > DRV_1$, and in other parts, $DRV_0 < DRV_1$. This is clearly illustrated by Figure 8.15, let us see how. First, we extract the direction in the parameter space that has maximum influence on DRV_0. This direction is shown as $w_{1,DRV0}$ in the two dimensional parameter space example of Fig. 8.14. As we move along this vector in the statistical parameter space, the corresponding DRV_0 increases from low to high (from bad to good).[2] Other directions would show changes in DRV_0 that are no greater than the change along this direction. The figure plots the simulated DRV_0 and DRV_1 values from a 1,000-point Monte Carlo run, along this direction. It is clear that the two DRV measures are inversely related: one decreases as the other increases.

Now, let us take the maximum as in (8.20), and choose the classification threshold t_c equal to the 97th percentile. Then, we pick out the worst 3% points from the classifier training data and plot them against the same latent variable in Fig. 8.15, as squares. Note that we have not trained the classifier yet, we are just looking at the points that the classifier would have to classify as being in the tail. We can clearly see that these points (squares) lie in two disjoint parts of the parameter space.

Since the true tail region defined by the tail threshold $t > t_c$ will be a subset of the classifier tail region (defined by t_c), it is obvious that the true tail region consists of two disjoint regions of the parameter space. This is illustrated in our two dimensional example in Fig. 8.14. The dark tail regions on the top-right and bottom-left corners of the parameter space correspond to the large DRV values shown as squares in Fig. 8.15.

8.5.4.1 The Solution

Instead of building a single classifier for the tail of in (8.20), let us build two separate classifiers, one for the 97th percentile (t_{c,DRV_0}) of DRV_0, and another for

---

[2]Here, we extract this vector using the SiLVR tool of [20] so that $w_{1,DRV0}$ is essentially the projection vector of the first latent variable of DRV_0

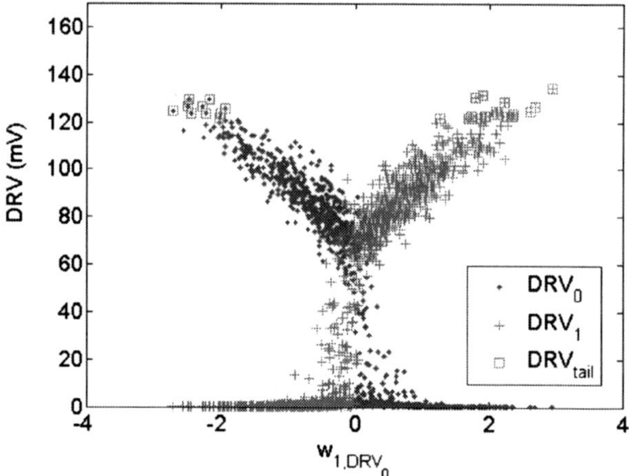

Fig. 8.15 Behavior of DRV_0 and DRV_1 along the direction of maximum variation in DRV_0. The worst 3% DRV values are plotted as *squares*, showing the disjoint tail regions (along this direction in the parameter space.)

the 97th percentile (t_{c, DRV_1}) of DRV_1. The generated Monte Carlo samples can then be filtered through both these classifiers: points classified as "body" by *both* the classifiers will be blocked, and the rest will be simulated. In the general case for arbitrary number of arguments in the conditional, let the circuit metric be given as

$$y = \max(y_0, y_1, \ldots). \tag{8.21}$$

Note that a min( ) operator can easily be converted to max( ) by negating the arguments. The resulting general algorithm is then as follows:

1. Perform initial sampling to generate training data to build the classifiers, and estimate tail and classification thresholds, t_i and $t_{c,i}$, respectively, for each y_i, $i = 0, 1, \ldots$. Also estimate the tail threshold t for y.
2. For each argument, y_i, $i = 0, 1, \ldots$, of the conditional (8.21), build a classifier C_i at a classification threshold $t_{c,i}$ that is less than the corresponding tail threshold t_i.
3. Generate more points using Monte Carlo, but block the points classified as "body" by *all* the classifiers. Simulate the rest and compute y for the simulated points.

Hence, in the case of Fig. 8.14, we build a separate classifier for each of the two boundaries. The resulting classification boundaries are shown as solid lines. From the resulting simulated points, those with $y > t$ are chosen as tail points for further analysis, e.g., for computing a GPD model for the tail distribution of y.

Note that this same algorithm can also be used for the case of multiple circuit metrics. Each metric would have its own thresholds and its own classifier, just like each argument in (8.21), the only difference being that we would not be computing any conditional.

Note also that adding additional classifiers increases the number of points that require simulation: if two nonoverlapping classifiers each identify 3% of points, then overall 6% of points will need simulation.

8.5.5   Extremely Rare Events and Their Statistics

The GPD tail model can be used to make predictions regarding rare events that are farther out in the tail than any of the data we used to compute the GPD model. Indeed, this is the compelling reason for adopting the GPD model. However, as suggested by common intuition and the estimate variance relations of Sect. 8.4, we expect the statistical confidence in the estimates to decrease as we predict farther out in the tail. Let us run an experiment to test this expectation.

Let us go back to our SRAM cell write time example. Suppose we run 50 runs of Monte Carlo with $n_{MC} = 100,000$ points each and compute a GPD tail model from each run, using tail threshold $t =$ the 99th percentile write time. This gives us 50 slightly different pairs of the GPD parameters (ξ, β), one for each of 50 GPD models so computed, and 50 different estimates of the $m\sigma$ point, where $m \in [3,6]$. These estimates are shown in Fig. 8.16.

As expected, the spread of the estimates increases as we extrapolate further with the GPD model. The general inference is that we should not rely on the GPD tail model too far out from our data. How then do we compute a more reliable GPD model further out in the tail?

A solution is to sample further out in the tail and use a higher tail threshold for building the GPD model of the tail. This is, of course, "easier said than done." Suppose we wish to support our GPD model with data up to the 6σ point. The failure probability of a 6σ value is roughly 1 part per *billion*, corresponding to a 99% chip yield requirement for a 10 Mb cache.

This is definitely not an impractical requirement. However, for a 99% tail threshold, even a perfect classifier ($t_c = t$) will only reduce the number of simulations to an extremely large 10 million. If we decide to use a 99.9999% threshold, the number of simulations will be reduced to a more practical 1,000 tail points (with a

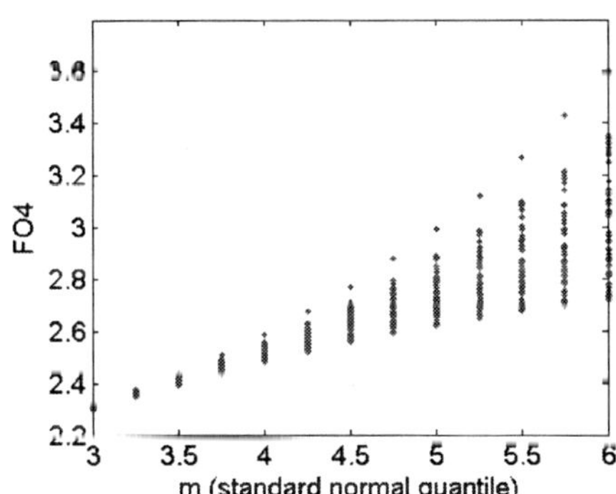

Fig. 8.16 The spread of $m\sigma$ point estimates across 50 runs of statistical blockade

perfect classifier). However, we will need to simulate an extremely large number of points (≥ 1 million) to generate a classifier training set with at least one point in the tail region. In both cases, the circuit simulation counts are too high. We now describe a recursive formulation of statistical blockade from [27] that reduces this count drastically.

8.5.6 A Recursive Formulation of Statistical Blockade

Let us first assume that there are no conditionals. For a tail threshold equal to the ath percentile, let us represent it as t^a, and the corresponding classification threshold as t_c^a. For this threshold, build a classifier C^a and generate sufficient points beyond the tail threshold, $y > t^a$, so that a *higher* percentile (t^b, t_c^b, $b > a$) can be estimated. For this new, higher threshold (t_c^b, a new classifier C^b is trained and a new set of tail points ($y > t^b$)) is generated. This new classifier will block many more points than C^a, significantly reducing the number of simulations. This procedure is repeated to push the threshold out more until the tail region of interest is reached.

The complete algorithm, applied to general conditionals, is shown in Algorithm 8.2. This pseudocode uses a *conditional* tail threshold of the 99th percentile at each recursion stage: (t^a, t_c^a) = (99%, 97%) points, (t^b, t_c^b) = (99.99%, 99.97%) points, and so on. Consequently, the total sample size (without filtering) n is restricted to some power of 100, times 1,000:

$$n = 100^j \cdot 1000, \quad j = 0, 1, \ldots \tag{8.22}$$

These threshold values are an extension of the values chosen in the original statistical blockade paper [21], where the authors relied on empirical evidence. A general version of the algorithm for arbitrary thresholds is presented in [6, 22]. Practical implementation of this general algorithm is, however, difficult and is a topic for further research.

The functions that appear also in Algorithm 8.1 do the same work here, hence, we do not reiterate their description. $\mathbf{f}_{sim}$ now returns multiple outputs; it computes the values of all the arguments of the conditional in $y = \max(y_0, y_1, \ldots)$. For example, in the case of DRV, it returns the values of DRV_0 and DRV_1. These values, for any one Monte Carlo point, are stored in one row of the result matrix $\mathbf{Y}$. The function $MonteCarloNext(\Delta n)$ returns the *next* Δn points in the sequence of points generated till now. The function $GetWorst(n, \mathbf{X}, \mathbf{y})$ returns the n worst values in the vector $\mathbf{y}$ and the corresponding rows of the matrix $\mathbf{X}$. This functionality naturally extends to the two argument $GetWorst(n, \mathbf{y})$. $GetGreaterThan(t, \mathbf{X}, \mathbf{y})$ returns the elements of $\mathbf{y}$ that are greater than t, along with the corresponding rows of $\mathbf{X}$.

Require: initial sample size n_0 (e.g., 1,000);

total sample size n; tail sample size n_t;

performance metric function $y = \max(y_0, y_1, \ldots)$

1. $\mathbf{X} = \text{MonteCarlo}(n_0)$
2. $n' = n_0$
4. $\mathbf{Y} = \mathbf{f}_{\text{sim}}(\mathbf{X})$ // Simulate the initial Monte Carlo sample set
5. $\mathbf{y}_{\text{tail},i} = \mathbf{Y}_{\bullet,i}$, $i = 0, 1, \ldots$ // The i-th column of $\mathbf{Y}$ contains values for y_i in $y = \max(y_0, y_1, \ldots)$
6. $\mathbf{X}_{\text{tail},i} = \mathbf{X}$, $i = 0, 1, \ldots$
7. **while** $n' < n$
8. $\Delta n = 99n'$ // Number of points to filter in this recursion step
10. $n' = n' + \Delta n$ // Total number of points filtered by the end of this recursion stage
11. $\mathbf{X} = \text{MonteCarloNext}(\Delta n)$ // The next Δn points in the Monte Carlo sequence
12.   **forall** i : y_i is an argument in $y = \max(y_0, y_1, \ldots)$
13. $(\mathbf{X}_{\text{tail},i}, \mathbf{y}_{\text{tail},i}) = \text{GetWorst}(1{,}000, \mathbf{X}_{\text{tail},i}, \mathbf{y}_{\text{tail},i})$ // Get the 1,000 worst points
14. $t = \text{Percentile}(\mathbf{y}_{\text{tail},i}, 99)$
15. $t_c = \text{Percentile}(\mathbf{y}_{\text{tail},i}, 97)$
16. $C^i = \text{BuildClassifier}(\mathbf{X}_{\text{tail},i}, \mathbf{y}_{\text{tail},i}, t_c)$
17. $(\mathbf{X}_{\text{tail},i}, \mathbf{y}_{\text{tail},i}) = \text{GetGreaterThan}(t, \mathbf{X}_{\text{tail},i}, \mathbf{y}_{\text{tail},i})$ // Get the points with $y_i > t$
18. $\mathbf{X}_{\text{cand},i} = \text{Filter}(C^i, \mathbf{X})$ // Candidate tail points for y_i
19.   **endfor**
20. $\mathbf{X} = \begin{bmatrix} \mathbf{X}_{\text{cand},0}^T & \mathbf{X}_{\text{cand},1}^T & \ldots \end{bmatrix}^T$ // Union of *all* candidate tail points
21. $\mathbf{Y} = \mathbf{f}_{\text{sim}}(\mathbf{X})$ // Simulate all candidate tail points
22. $\mathbf{y}_{\text{cand},i} = \{\mathbf{Y}_{j,i} : \mathbf{X}_{j,\bullet} \in \mathbf{X}_{\text{cand},i}\}$, $i = 0, 1, \ldots$ // Extract the tail points for y_i
23. $\mathbf{y}_{\text{tail},i} = \begin{bmatrix} \mathbf{y}_{\text{tail},i}^T & \mathbf{y}_{\text{cand},i}^T \end{bmatrix}^T$, $\mathbf{X}_{\text{tail},i} = \begin{bmatrix} \mathbf{X}_{\text{tail},i}^T & \mathbf{X}_{\text{cand},i}^T \end{bmatrix}^T$, $i = 0, 1, \ldots$ // All tail points till now
24. **endwhile**
25. $\mathbf{y}_{\text{tail}} = \text{MaxOverRows}([\mathbf{y}_{\text{tail},0} \;\; \mathbf{y}_{\text{tail},1} \;\ldots])$ // compute the conditional
26. $\mathbf{y}_{\text{tail}} = \text{GetWorst}(n_t, \mathbf{y}_{\text{tail}})$
27. $(\xi, \beta) = \text{FitGPD}(\mathbf{y}_{\text{tail}} - \min(\mathbf{y}_{\text{tail}}))$

Algorithm 8.2. *The recursive statistical blockade algorithm with fixed sequences for the tail and classification thresholds:* t $= 99\%, 99.99\%, 99.9999\%, \ldots$ *points, and* t$_c = 97\%, 99.97\%, 99.9997\%, \ldots$ *points. The total sample size is given by* (8.22).

The algorithms presented here are in iterative form, rather than recursive form. To see how the recursion works, suppose we want to estimate the 99.9999% tail. To generate points at and beyond this threshold, we first estimate the 99.99% point and use a classifier at the 99.97% point to generate these points efficiently. To build this classifier in turn, we first estimate the 99% point and use a classifier at the 97% point. Figure 8.17 illustrates this recursion on the PDF of any one argument in the conditional (8.21).

8.5.6.1 An Experiment with Data Retention Voltage

We now test the recursive statistical blockade method on another SRAM cell test case, where we compute DRV as in (8.20). In this case, the SRAM cell is implemented in an industrial 90 nm process. Wang et al. [23] develop an analytical model

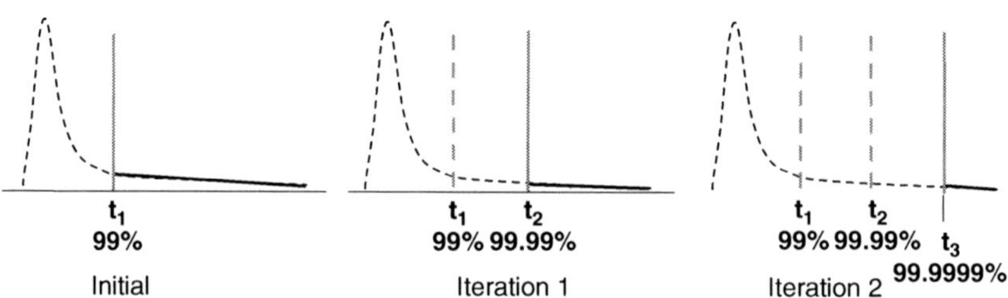

Fig. 8.17 Recursive formulation of statistical blockade as in Algorithm 8.2

for predicting the CDF of the DRV that uses not more than 5,000 Monte Carlo points. The CDF is given as

$$F(y) = 1 - \mathrm{erfc}(y_0) + \frac{1}{4}\mathrm{erfc}^2(y_0), \quad \text{where } y_0 = \frac{\mu_0 + k(y - V_0)}{\sqrt{2}\sigma_0}, \qquad (8.23)$$

where y is the DRV value and erfc( ) is the complementary error function [24]. k is the sensitivity of the SNM of the SRAM cell to the supply voltage, computed using a DC sweep. μ_0 and σ_0 are the mean and standard deviation of the SNM (SNM$_0$), for a user-defined supply voltage V_0. SNM$_0$ is the SNM of the cell while storing a 0. These statistics are computed using a short Monte Carlo run of 1,500–5,000 sample points. We direct the reader to [23] for complete details regarding this analytical model of the DRV distribution. The qth quantile can be estimated as

$$\mathrm{DRV}(q) = \frac{1}{k}\left(\sqrt{2}\sigma_0\mathrm{erfc}^{-1}(2 - 2\sqrt{q}) - \mu_0\right) + V_0. \qquad (8.24)$$

Here, DRV(q) is the supply voltage V_{dd} such that $P(\mathrm{DRV}(q)) \leq V_{dd} = q$.

Let us now compute the DRV quantiles as $m\sigma$ points, such that q is the cumulative probability for the value m from a standard normal distribution. We use five different methods to estimate the DRV quantiles for $m \in [3,8]$:

1. *Analytical*: Use equation (8.24).
2. *Recursive statistical blockade without the GPD model*: Algorithm 8.2 is run for $n = 1$ billion. This results in three recursion stages, corresponding to total sample sizes of $n' = 100,000$, 10 million and 1 billion Monte Carlo points, respectively. The worst DRV value for these three recursion stages are estimates of the 4.26σ, 5.2σ, and 6σ points, respectively.
3. *GPD model from recursive statistical blockade*: The 1,000 tail points from the last recursion stage of the recursive statistical blockade run are used to fit a GPD model, which is then used to predict the DRV quantiles.
4. *Normal*: A normal distribution is fit to data from a 1,000 point Monte Carlo run, and used to predict the DRV quantiles.
5. *Lognormal*: A lognormal distribution is fit to the same set of 1,000 Monte Carlo points, and used for the predictions.

The results are shown in Fig. 8.18. From the plots in the figure, we can immediately see that the recursive statistical blockade estimates are very close to the estimates from the analytical model. This shows the efficiency of the recursive formulation in reducing the error in predictions for events far out in the tail.

Table 8.3 shows the number of circuit simulations performed at each recursion stage. The total number of circuit simulations is 41,721. This is not small, but in comparison to standard Monte Carlo (1 billion simulations), and basic, nonrecursive statistical blockade (approximately, 30 million with $t_c = 97$th percentile) it is extremely fast. 41,721 simulations for DRV computation of a 6 T SRAM cell can be completed in several hours on a single computer today. With the advent of multicore processors, the total simulation time can be drastically reduced with proper implementation.

Note that we can extend the prediction power to 8σ with the GPD model, without any additional simulations. Standard Monte Carlo would need over 1.5 quadrillion circuit simulations to generate a single 8σ point. For this case, the speedup over standard Monte Carlo is extremely large. As expected, the normal and lognormal fits show large errors. The normal fit is unable to capture the skewness of the DRV distribution. On the other hand, the lognormal distribution has a heavier tail than the DRV distribution.

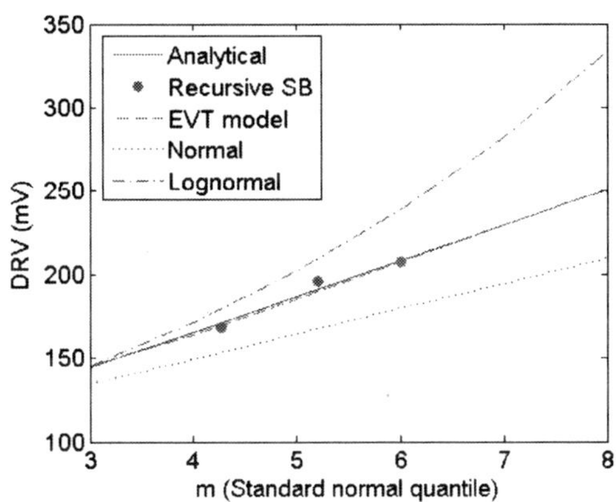

Fig. 8.18 Estimates of DRV quantiles from five estimation methods. The GPD model closely fits the analytical model (8.23). The *solid circles* show the worst DRV values from the three recursion stages of statistical blockade sampling. The normal and lognormal models are quite inaccurate

Table 8.3 Number of circuit simulation needed by recursive statistical blockade to generate a 6σ point

Recursion stage	Number of simulations
Initial	1,000
1	11,032
2	14,184
3	15,505
Total	41,721
Speedup over Monte Carlo	23,969×
Speedup over statistical blockade	719×

8.6   Combining Effects Across a Memory

The preceding sections have shown how EVT can be used to calculate failure probabilities of a property such as DRV that applies to a single bit cell. Expanding DRV to an entire memory is simple: the minimum DRV across all bit cells is the DRV for the memory as a whole. If repair is available, then to a first approximation n cells can be repaired with n repairable elements, leaving the $(n + 1)$th worst-case value. Many properties, however, do not generalize so easily. This section explores how EVT may be applied to systems comprised of multiple interacting components. It uses the example of slow reads, caused by a combination of weak bit cell read current, sense amp device mismatch, and self-timing path variation. The description follows that of [25].

We consider this problem in three steps: First, we consider a key subsystem in a memory: the individual bit cells associated with a given sense amp. Next, we generalize these results to the entire memory. Finally, we extend these results to multiple instances on a chip. For simplicity, we assume a Gaussian distribution of the underlying values. Actual read currents, for example, follow a slightly skewed distribution (high currents are more common than low currents), requiring the method here to be adapted if higher accuracy is desired. Recall from Table 8.1 that Gaussian data leads to a Gumbel distribution of worst case values.

8.6.1   Subsystem Results: Bit Cells Connected to a Given Sense Amplifier

Figure 8.19 shows the distribution of worst-case bit cell read current values from Monte Carlo simulation (2,048 trials where the worst case of 4,096 samples was taken). Note that the distribution shown in the figure is very similar to the Gumbel curves in Fig. 8.20 – right skewed, variable width.

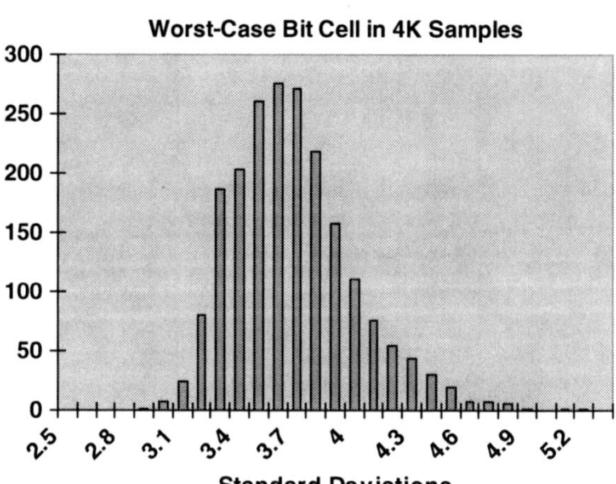

Fig. 8.19 Distribution giving standard deviations away from the mean of 2,048 instances of the worst-case bit cell among 4,096 random trials

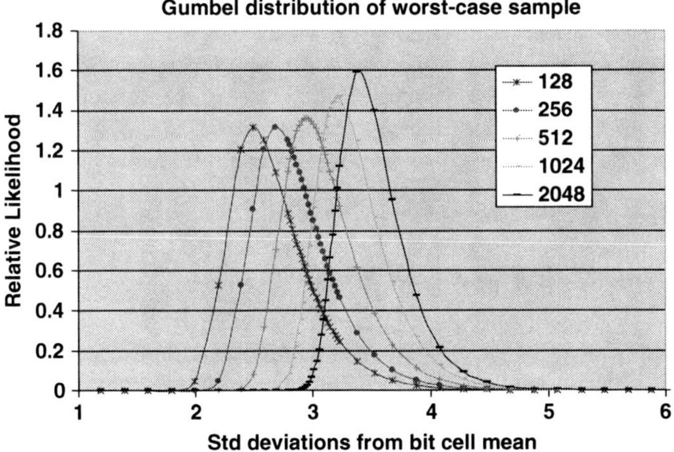

Fig. 8.20 Gumbel distributions for various sample sizes

Given this, we may now attempt to identify the correct Gumbel parameters for a given memory setup, using the procedure below, applied to a system M0 consisting of a single sense amp and N bit cells, where each bit cell has mean read current μ and standard deviation σ. Note that the CDF of the Gumbel distribution of mode u and standard deviation s is given by

$$\mathrm{CDFG}(y) = \int_{-\infty}^{y} \frac{1}{s}\exp\left[\frac{-(x-u)}{s} - e^{\frac{-(x-u)}{s}}\right]dx \qquad (8.25)$$

Begin with an estimate for a Gumbel distribution that matches the appropriate distribution shown in Fig. 8.19 – the value for u should be slightly less than the 50th percentile point, and a good starting point for s is 0.25 (*see* Fig. 8.20).

1. Given an estimated u and s, a range of CDFG(y) values can be calculated. We use numerical integration taking many sample points around the mode value u and a more fewer elsewhere. For example, with $u = 3.22$ and $s = 0.25$, the CDFG calculated for $y = 3.40$ is 0.710.
2. Equation 8.25 allows us to calculate the cumulative distribution function R for a given probability p and sample number N. Thus, for each of the CDFG values arrived at in step 2, it is possible to calculate R. Continuing the previous example, R(1024,0.710) = 3.402.
3. Repeat steps 2–3, adjusting u and s as necessary in order to minimize the least squares deviation of y and R(N, CDFG(y)) across the selected set y.

Using this approach results in the following Gumbel values for N of interest in margining the combination of sense amp and bit cells:

These values produce the Gumbel distributions shown in Fig. 8.20.

Recall that these distributions show the probability distribution for the worst-case bit cell for sample sizes ranging from 128 to 2048. From these curves, it is quite clear that margining the bit cell and sense amp subsystem (M0) to three sigma of bit cell current variation is unacceptable for any of the memory configurations. Five sigma, on the other hand, appears to cover essentially all of the distributions. This covers M0, but now needs to be extended to a full memory.

8.6.2 Generalizing to a Full Memory

Several methods can be used alone or in combination to validate that a memory design has sufficient read margin. One common approach is to insert an artificial voltage offset across the bit lines as shown in Fig. 8.11. The primary purpose of this offset is to represent the mismatch in threshold between the pass transistors of the sense amplifier, but the same method can be used to compensate for other effects as well. This section shows how the offset method can be used to cover both sense amplifier and bit cell variability.

Suppose that a 512 Kbit memory (system M1, see for example Fig. 8.21) consists of 512 copies of a subsystem (M0) of one sense amp and 1,024 bit cells. How should this be margined? Intuitively, the bit cells should have more influence than the sense amps, since there are more of them, but how can this be quantified?

There are various possible approaches, but we consider two representative methods: isolated and merged.

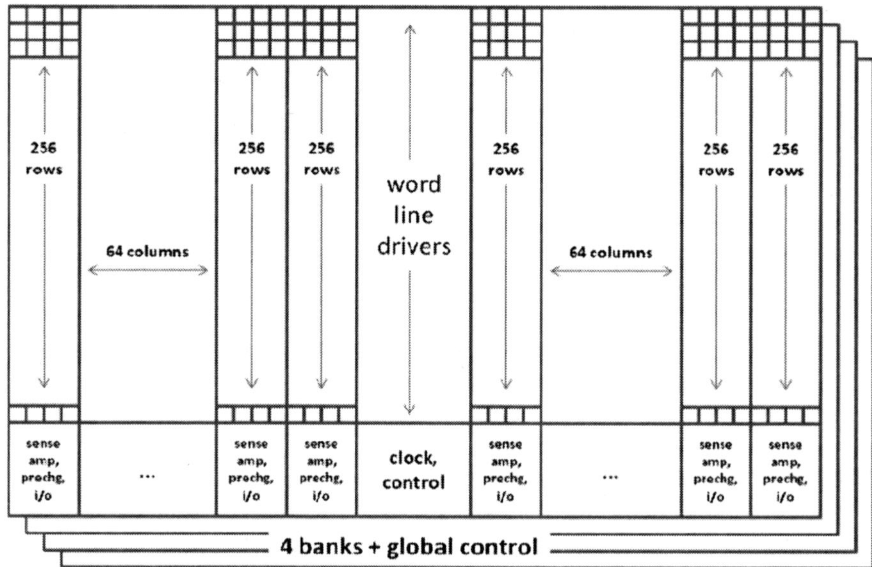

Fig. 8.21 Example of a 512 Kbit memory composed of four banks of 1,024 words, each 128 bits wide, using a four column multiplexing structure

8.6.2.1   Method 1 (Isolated)

In this approach, choose a high value for the CDF of the subsystem M0 (e.g., set up the sense amp to tolerate the worst-case bit cell with probability 99.99%) and treat this as independent from system M1.

8.6.2.2   Method 2 (Merged)

Attempt to merge effects of the subsystem M0 into the larger M1 by numerically combining the distributions, i.e,. finding the worst-case combination of bit cell and sense amp within the system, being aware that the worst-case bit cell within the entire memory is unlikely to be associated with the worst-case sense amp, as shown graphically in Fig. 8.22.

Each method has some associated issues. For method 1, there will be many "outside of margin" cases (e.g., a bit cell just outside the 99.99% threshold with a better than average sense amp) where it cannot be proven that the system will pass, but where in all likelihood it will. Thus, the margin method will be unduly pessimistic. On the other hand, combining the distributions also forces a combined margin technique (e.g., a sense amp offset to compensate for bit cell variation) which requires that the relative contributions of both effects to be calculated prior to setting up the margin check.

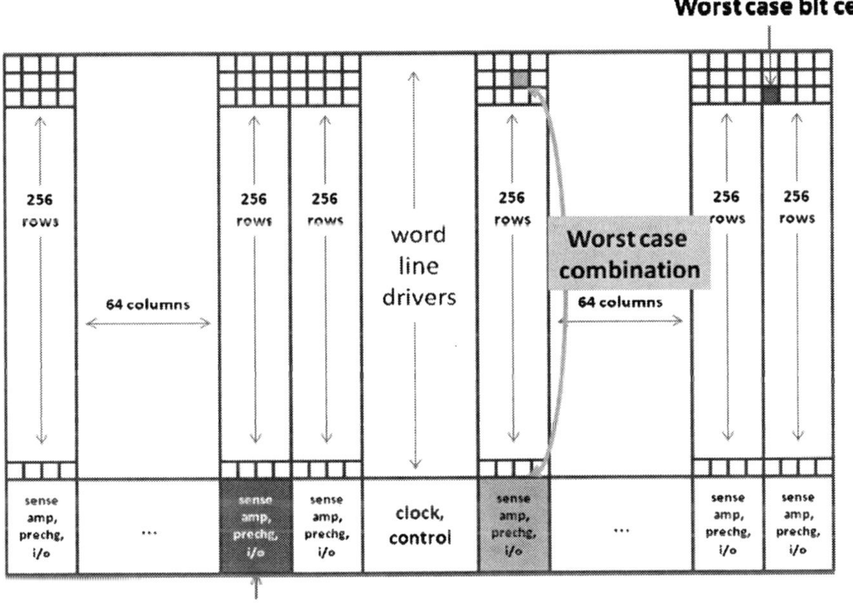

Fig. 8.22 Example showing worst-case bit cell, sense amp, and combination in a memory

Consider systems M1 and M0 as described above. We are concerned with local variation within these systems (mainly implant), and, as with Monte Carlo SPICE simulation, we can assume that variations in bit cells and sense amps are independent. Since there are 1,024 bit cells in M0, we can use a Gumbel distribution with $x = 3.22$ and $s = 0.25$ (see Table 8.4). Similarly, with 512 sense amps in system M1, we can use a Gumbel distribution with $x = 2.95$ and $s = 0.27$. We can plot the joint distribution of worst-case bit cell and worst-case sense amplifier of this combined in three dimensions as shown in Fig. 8.23. The skewed nature of both distributions can be seen, as well as the combined probability, which shows that the *most likely* worst-case combination is, as expected, a sense amp 2.95 sigma from the mean together with a bit cell 3.22 sigma from the mean.

Margin methodologies can be considered graphically by looking on Fig. 8.23 from above. Method 1 consists of drawing two lines, one for bit cells and one for sense amps, cutting the space into quadrants, as shown in Fig. 8.24, set to (for e-xample) 5 sigma for both bit cells and sense amps. Only the lower left quadrant is guaranteed to be covered by the margin methodology; these designs have both sense amp and bit cell within the margined range. Various approaches to

Table 8.4 Gumbel parameters for various sample sizes

N	128	256	512	1,024	2,048	4,096
Mode	2.5	2.7	2.95	3.22	3.4	3.57
Scale	0.28	0.28	0.27	0.25	0.23	0.21

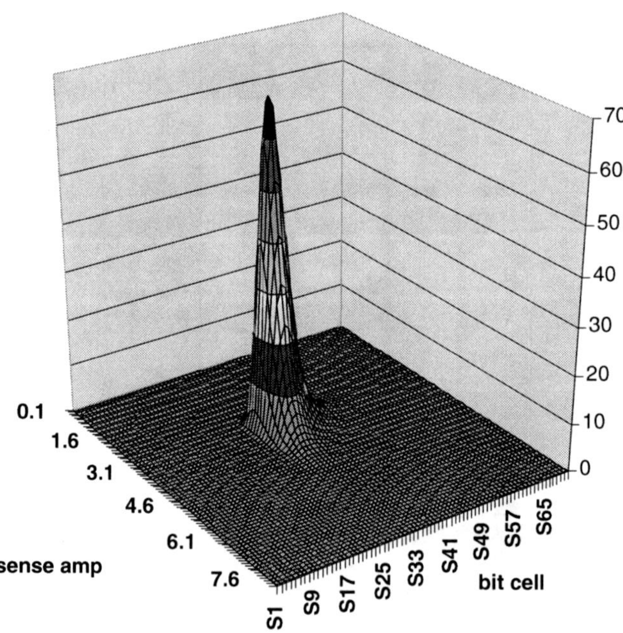

Fig. 8.23 Combined Gumbel distribution

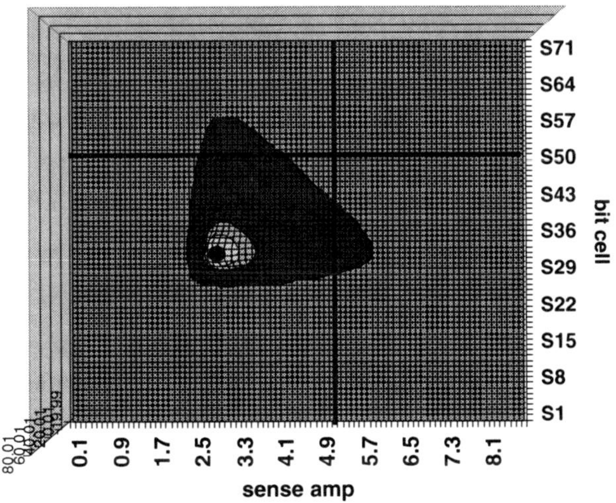

Fig. 8.24 Margin method 1, separate margining of bit cell and sense amplifier

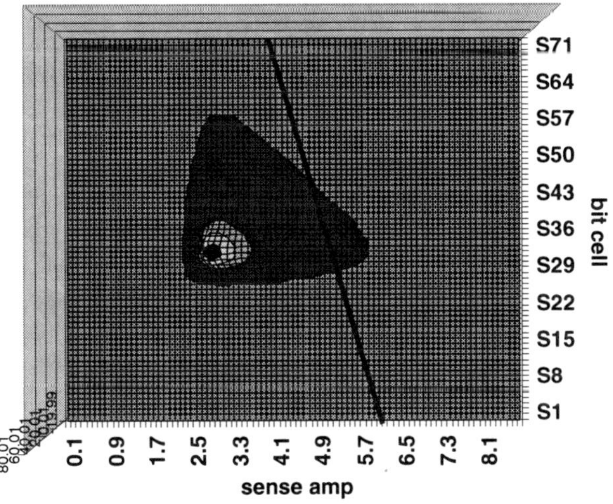

Fig. 8.25 Margin method 2, combined margining of bit cell and sense amplifier

method 2 are possible, but one involves converting bit cell current variation to a sense amp voltage offset via Monte Carlo simulation. For example, if expected bit line voltage separation when the sense amp is latched is 100 mV and Monte Carlo simulation shows that a 3 sigma bit cell produces only 70 mV of separation, then 1 sigma of bit cell read current variation is equivalent to 10 mV of bit line differential. This method is represented graphically by an angled line, as shown in Fig. 8.25. Regions to the left of the line are combinations of bit cell and sense amp within margin range.

Using numerical integration methods, the relative coverage of the margin approaches can be compared. Method 1 covers 99.8% of the total distribution (roughly 0.1% in each of the top left and lower right quadrants). Method 2 covers 99.9% of the total distribution, and is easier for a design to meet because it passes closer to the peak of the distribution (a margin line that passes through the intersection of the two lines in Fig. 8.24 requires an additional 0.5 sigma of bit cell offset in this case).

A complete memory, as in Fig. 8.3, has an additional layer of complexity beyond what has been considered so far (e.g., self-timing path, clocking circuitry), which we denote as system M2. There is only one copy of M2, and so a Gaussian distribution will suffice for it. It is difficult to represent this effect graphically, but it too can be compensated for by adding offset to a sense amplifier.

Extension to repairable memory: Note that the existence of repairable elements does not alter these calculations much. System M0 cannot in most architectures be improved by repair, but system M1 can. However, repair is primarily dedicated to defects, and so the number of repairs that can be allocated to out-of-margin design is usually small.

8.6.3 Generalization to Multiple Memories

Since failure of a single memory can cause the entire chip to fail, the margin calculation is simply the combined probability that each memory is within margin (hence the product of the individual margin probabilities). For a chip with one copy of the memory in the previous section, this probability is 99.9% under Method 2. This drops to 95% for 50 memories and to 37% for 1,000 memories. The latter two numbers are significant. If 50 memories are typical, then this approach leads to a margin method that will cover all memories on a typical chip 95% of the time, and so variability-related yield will be at least 95%. This is not ideal, but perhaps adequate. On the other hand, if 1,000 memories are typical, the majority of chips will contain at least 1 memory that is outside its margin range, and the lower bound on variability related yield is only 37%. This is clearly unacceptable. To achieve a 99% lower bound for yield on 1,000 memories, each individual yield must be 99.999%, which is outside the shaded area in Fig. 8.25. For this reason, it is vital to consider the entire chip when developing a margin method.

8.7 Conclusions

As we have seen in this chapter, the application of EVT to memory statistics is a fertile research area. As devices scale, variability challenges increase, as do demands for increased performance and reduced power. The sheer number of nearly identical devices in memory means that margins must be calculated for very rare events, and that the tails of distributions must be well understood, not just for bit cells, but for common peripheral cells as well. Future work can be expected to focus on quantifying these tails as well as pushing designs further in terms of variation tolerance, variation immunity, and variation exploitation – working with variation in order to push bounds on achievable performance. Most of the techniques described in this chapter are applicable to nontraditional forms of memory as well, and as such will continue to be studied for many years to come.

References

1. Kwai D-M et al (2000) Detection of SRAM cell stability by lowering array supply voltage, Proc Asian Test Symp, 2000
2. Embrechts P, Klüppelberg C, Mikosch T (2003) Modelling extremal events for insurance and finance, 4th edn. Springer-Verlag, Berlin
3. Fisher RA, Tippett LHC (1928) Limiting forms of the frequency distribution of the largest or smallest member of a sample, Proc Cambridge Philos Soc 24:180–190
4. Gnedenko B (1943) Sur la distribution limite du terme maximum d'une aleatoire. Ann Math 44(3):423–453
5. Resnick SI (1987) Extreme values, regular variation and point processes. Springer, New York
6. Singhee A (2007) Novel algorithms for fast statistical analysis of scaled circuits, PhD thesis, Electrical and Computer Engg., Carnegie Mellon University, 2007
7. Balkema AA, de Haan L (1974) Residual life time at great age. Ann Prob 2(5):792–804
8. Pickands J III (1975) Statistical inference using extreme order statistics. Ann Stats 3(1):119–131
9. Davison AC, Smith RL (1990) Models for exceedances over high thresholds (with discussion). J Royal Stat Soc B 52:393–442
10. Grimshaw SD (1993) Computing maximum likelihood estimates for the generalized Pareto distribution. Technometrics 35(2):185–191
11. Smith RL (1987) Estimating tails of probability distributions. Ann Stats 15(3):1174–1207
12. Smith RL (1985) Maximum likelihood estimation in a class of non-regular cases. Biometrika 72:67–92
13. Hosking JRM (1986) The theory of probability weighted moments, IBM Research Report, RC12210, 1986
14. Hosking JRM, Wallis JR (1987) Parameter and quantile estimation for the generalized Pareto distribution. Technometrics 29(3):339–349
15. Hastie T, Tibshirani R, Friedman J (2001) The elements of statistical learning: data mining, inference, and prediction, Springer, 2001
16. Morik K, Brockhausen P, Joachims T (1999) Combining statistical learning with a knowledge-based approach – a case study in intensive care monitoring, Proc 16th Intn'l Conf Machine Learning, 1999
17. Liu W, Jin X, Chen J, Jeng M-C, Liu Z, Cheng Y, Chen K, Chan M, Hui K, Huang J, Tu R, Ko P, Hu C (1988) BSIM 3v3.2 Mosfet model users' manual, Univ California, Berkeley, Tech. Report No. UCB/ERL M98/51, 1988
18. Joachims T (1999) Making large-scale SVM learning practical. In Schölkopf B, Burges C, Smola A (eds) Advances in Kernel methods – support vector learning. MIT Press, 1999
19. Krishnamurthy RK, Alvandpour A, De V, Borkar S (2002) High-performance and low-power challenges for sub 70 nm microprocessor circuits, Proc Custom Integ Circ Conf, 2002
20. Singhee A, Rutenbar RA (2007) Beyond low order statistical response surfaces: latent variable regression for efficient, highly nonlinear fitting. Proc IEEE/ACM Design Autom Conf, 2007

21. Singhee A, Rutenbar RA (2007) Statistical blockade: a novel method for very fast Monte Carlo simulation of rare circuit events, and its application, Proc Design Autom Test Europe, 2007
22. Zhang K (2009) Embedded memories for nanoscale VLSIs, Springer, 2009
23. Wang J, Singhee A, Rutenbar RA, Calhoun BH (2007) Modeling the minimum standby supply voltage of a full SRAM array. Proc Europ Solid State Cir Conf, 2007
24. Press WH, Flannery BP, Teukolsky AA, Vetterling WT (1992) Numerical recipes in C: the art of scientific computing, Cambridge University Press, 2nd ed, 1992
25. Aitken R, Idgunji S (2007) Worst-case design and margin for embedded SRAM. Proc Design Autom Test Europe, 2007
26. Aitken RC (2007) Design for manufacturability and yield enhancement. In Wang LT (ed) System-on-chip test architecture: nanometer design for tesability, Elsevier
27. Singhee A, Wang J, Calhoun BH, Rutenbar RA (2008) Recursive Statistical Blockade: an enhanced technique for rare event simulation with application to SRAM circuit design, Proc Int Conf VLSI Design, 2008

Index

CPSIA information can be obtained at www.ICGtesting.com
Printed in the USA
238335LV00005B/34/P

9 781441 966056